ECONOMICS, ETHICS, AND
ENVIRONMENTAL POLICY

"This collection of essays, deeply informed in economic theory and in ethical analysis, provides the best current exploration of the intellectual hinterland between them. The authors, both philosophers and economists, fully understand each other's positions and objections, and so this volume takes a giant step in clarifying and resolving disagreements that have long vexed environmental policy making." *Mark Sagoff, University of Maryland*

D0814457

Economics, Ethics, and Environmental Policy

Contested Choices

EDITED BY

Daniel W. Bromley
and
Jouni Paavola

Blackwell Publishing

© 2002 by Blackwell Publishers Ltd
a Blackwell Publishing company
except for editorial arrangement and introduction
© 2002 by Daniel W. Bromley and Jouni Paavola

Editorial Offices:
108 Cowley Road, Oxford OX4 1JF, UK
Tel: +44 (0)1865 791100
350 Main Street, Malden, MA 02148-5018, USA
Tel: +1 781 388 8250

First published 2002 by Blackwell Publishers Ltd

Library of Congress Cataloging-in-Publication Data has been applied for

ISBN 0–631–22968–X (hardback); 0–631–22969–8 (paperback);

A catalogue record for this title is available from the British Library.

Set in 10/12pt Meridien
by Graphicraft Limited, Hong Kong

For further information on
Blackwell Publishers, visit our website:
www.blackwellpublishers.co.uk

Contents

Figures

Tables

Contributors

Daniel W. Bromley is Anderson–Bascom Professor of Applied Economics at the Department of Agricultural and Applied Economics at the University of Wisconsin–Madison, USA.

Nick Hanley is a professor at the Economics Department of the University of Glasgow, UK.

Juha Hiedanpää is a researcher at the Satakunta Environmental Research Centre of the University of Turku, Finland.

Alan Holland is a Professor of Applied Philosophy at the Institute for Environment, Philosophy and Public Policy at Lancaster University, UK.

Olof Johansson-Stenman is an associate professor at the Department of Economics of Göteborg University, Sweden.

Nick Johnstone is an administrator at the National Policies Division of the Environment Directorate of OECD in Paris, France.

Munguti Katui-Katua is the Director of Community Management and Training Services (East Africa), Nairobi, Kenya.

Andreas Kontoleon is a doctoral candidate at the Department of Economics of University College London, UK.

Mark Mujwajuzi is a professor at the Institute of Resource Assessment, University of Dar es Salaam, Tanzania.

Bryan Norton is a professor at the School of Public Policy at the Georgia Institute of Technology, USA.

Martin O'Connor is Professor in Economic Science at the University of Versailles St-Quentin-en-Yvelines, France.

Jouni Paavola is a Senior Research Associate at the Centre for Social and Economic Research on the Global Environment (CSERGE) at the University of East Anglia and an Associate Fellow of the Oxford Centre for the Environment, Ethics, and Society (OCEES), Mansfield College, Oxford, UK.

Ina Porras is a visiting researcher at the Environmental Economics Programme, International Institute of Environment and Environment (IIED), London, UK.

Alan Randall is a professor at the Department of Agricultural, Environmental, and Development Economics at Ohio State University, USA.

A. Allan Schmid is University Distinguished Professor at the Department of Agricultural Economics at Michigan State University, USA.

Jason F. Shogren is Stroock Distinguished Professor of Natural Resource Conservation and Management at the Department of Economics and Finance at the University of Wyoming, USA.

Clive L. Spash is Research Professor of Environmental and Rural Economics at the Department of Agriculture and Forestry, University of Aberdeen, UK.

Timothy Swanson is a professor at the Department of Economics at University College London, UK.

John Thompson is the director of the Sustainable Agriculture and Rural Livelihoods Programme, International Institute of Environment and Environment (IIED), London, UK.

James Tumwine is a senior lecturer at the Department of Paediatrics and Child Health, Makerere Medical School, Kampala, Uganda.

Arild Vatn is a professor at the Department of Economics and Social Sciences at the Agricultural University of Norway.

Bhaskar Vira is a lecturer in the Department of Geography at the University of Cambridge, UK.

Elizabeth Wood is a research associate at the Environmental Economics Programme, International Institute of Environment and Environment (IIED), London, UK.

Preface and Acknowledgments

This volume brings together contributions from North American and European philosophers and economists who are interested in the ethical questions related to applying economic analysis to environmental issues. The theme of our volume is perhaps more pertinent today than ever before. The credibility of policy-makers and economists seems to have achieved a new low amidst the protests against the world trade system, the slow progress of international negotiations on the mitigation of climate change, and public health and environmental concerns related to mad cow disease and genetically modified organisms. We hope that this volume adds new insights into the proper role of economic analysis in these and other matters of public policy.

In the course of producing this volume we have become indebted to a number of people and organizations, to whom our expressions of gratitude are owed. First and foremost, we want to express our deep appreciation to Christopher and Anne Johnson. Their financial donation in the memory of their nephew William Robbins, grandson of the distinguished economist Lord Lionel Robbins, made it possible for the contributors to this volume to convene in Oxford at the William Robbins Memorial Seminar on Economics, Ethics and the Environment. The workshop took place at Mansfield College, Oxford, in March 2000. We also thank Christopher and Anne Johnson, Richard and Brenda Robbins, and Robert Nurick for their participation and many valuable contributions during the workshop. Thanks go also to the European Union's Environment Directorate General for its financial support to our workshop and the editing of this volume.

We also want to thank the "policy practitioners" who made such valuable contributions during the Workshop. They challenged us to think beyond the narrow academic confines of such familiarity to the rest of us. These individuals included: J. Clarence Davies from Resources for the Future in Washington, DC; Henry Derwent, Bob Dinwiddy, and Dinah Nichols from the UK Department of the Environment, Transport and the Regions; and Norman Glass from the UK Treasury. We sincerely hope we have been able to reciprocate their invaluable contributions to this volume by addressing at least some of the difficult issues they face as practitioners of – and participants in – environmental policy. We also present our sincere thanks to Neil Summerton, Director of OCEES and former Under-Secretary at the UK Department of the Environment, who perhaps better than anybody else could switch between the hats of an academic and a policy-maker.

Daniel W. Bromley
Jouni Paavola
April 2001

PART 1

Introduction

CHAPTER *1*

Contested Choices

Jouni Paavola and Daniel W. Bromley

We offer this volume in response to a growing unease with the way in which many economists – and the policy-makers informed by economic writings on environmental policy – address environmental issues. Occasionally, this dissatisfaction erupts into widely publicized media events. We have in mind here the aggressive demonstrations at the meeting of the World Trade Organization in Seattle in late 1999. These protests were aimed at the emerging regime of global trade, regarded by most economists as welfare-enhancing and thus inherently desirable. In France, there have been demonstrations against increased globalization epitomized by "le fast food." The concern there centers on the view that globalization threatens not only French livelihoods but also the broader cultural values of the French. Finally, demonstrations in The Hague during the latest international climate change policy negotiations directly challenged many economists' favored means – marketable pollution permits – of curbing greenhouse gas emissions.

These and other international media events have had their national counterparts. In the United Kingdom, the disposal of the Brent Spar oil drilling rig, and the use of genetically modified organisms, have been among the most provocative issues during the past decade. In Finland, discontent with intensive forest management practices, and weak protection of old-growth forests, have often turned environmental activists into tree-dwellers. In India, farmers have protested and burned genetically engineered crops. More recently, Germany, France, Italy, and Spain have been swept by anxiety over mad cow disease, in yet another demonstration that normal methods of risk assessment – a central feature of environmental economics – are not seen as legitimate by a large share of the general population.

Behind these highly visible media events, various environmental organizations, other activist groups – and indeed ordinary citizens – regularly engage in political action at the local, national, and international levels. These groups often advocate policy choices substantially at odds with the prescriptions of economists. They also oppose projects that have been deemed desirable by economists and/or policy-makers. The terms often used in this context, such as "NIMBY" (Not In My Back Yard) and "LULU" (Locally Unwanted Land Uses) capture but a part of the phenomenon, and certainly do not help us understand what is really going on.

So, *what is going on*? Are fundamentally sound economic prescriptions – and implied policy choices – being questioned by badly informed and misguided zealots who would otherwise readily accept the received wisdom if only they were able to understand the logic and rationale behind economic analysis and the collective choices prescribed by it? Should policy-makers simply ignore these protests? Or do standard economic approaches to collective action really have it wrong? Would public policy be less contentious if those in positions of authority simply ignored economics as a guide to policy-making on environmental (and perhaps other) issues? That is, should policy-makers ignore economists?

In this volume we hope to move beyond these simplistic questions. At the same time, we do believe that economists and policy-makers should take seriously the discontent with economic analysis and policy prescriptions – as well as with the collective choices informed by them. Ironically, when it concerns individual choice, economists are pleased to assume sound reasoning on the part of individual agents whom we vest with limitless cognitive capacities and perfect knowledge. But then economists often discredit their concerns in the face of our prescriptions when it comes to collective action. Can our inconsistency be laid solely at the feet of strategic behavior – and free riding – on the part of the citizenry when collective action is under consideration? There must be something else at work here.

We believe, in other words, that there remains much to learn about the logic of collective action that cannot be explained by traditional approaches to public policy. That is, rather than dismissing these environmental conflicts as the inevitable result of a citizenry insufficiently exposed to the impeccable logic and rigor of contemporary economics, we hope to demonstrate in these various contributions that there is something to be learned from collective expressions of concern and discontent over environmental conflicts. We will also suggest that trying to understand what goes on in the policy process provides fertile ground for future research at the intersection of economics and ethics. Indeed, we regard this volume as an important first step in the growing interest in work where philosophy and economics meet.

The reader will notice that many of the contributions here offer somewhat critical views of contemporary environmental economics. Yet, at another level, this volume presents a strong defense of environmental economics. We firmly believe that research at the intersection of environmental economics and ethics can improve the understanding of policy problems and choices. We also believe that this research can be helpful for policy-making on environmental and other issues. The kind of economics we envision may not be helpful in the conventional sense of identifying uniquely optimal solutions – a role that economics has never been able to perform (because it is impossible). We do believe, however, that the modified economic program we are striving for here can help interest groups involved in collective choices to revise their expectations, to respect different viewpoints about policy matters, and to facilitate the design of institutional arrangements that implement and realize agreed-upon collective choices.

Next, we will identify and briefly discuss those areas of environmental economics that seem to be most frequently contested, in order to provide a context for the contributions of this volume. The contributions will then be briefly described.

Criticisms of Environmental Economics

Three closely related but yet distinct practices are often contested when policy-makers seek to employ economics as a guide in making environmental policy choices. The first contested practice is the exclusive use of *welfare criteria* for analyzing and making choices concerning policies and projects that have environmental impacts. Second, *monetary valuation* of the environment by the use of contingent valuation methods (and other valuation methods) has encountered sharp criticism, just as it has become increasingly popular among environmental economists and policy-makers. Finally, the *discounting* of future benefits and costs when conducting benefit–cost analysis has remained a durable subject of criticism from environmentalists and others. In what follows, we will briefly address each contested area in greater detail.

Welfarism

The broadest and perhaps the most fundamental criticism of contemporary environmental economics is a response to its philosophical foundations. Environmental economics is founded upon a worldview in which independent and all-knowing individuals act upon their exclusively welfare-centered motivations. Therefore, the policy prescriptions of environmental economists seek to implement and realize the worldview contained in their assumptions. The anthropocentric, welfarist, and egoistic dimensions of environmental economics surface in different ways in contemporary policy debates and practices.

The conventional assumptions of environmental economics sometimes result in an insistence to set welfare-maximizing goals for environmental policy (see Palmer, Oates, and Portney, 1995; Arrow et al., 1996). A standard example is the idea of a socially optimal level of environmental harm – the level of an environmentally harmful activity at which the marginal (abatement) cost of reducing the harm equals the marginal benefit from the improvement of environmental quality. Economists will insist that this is the welfare-maximizing level of pollution or other environmental harm. One example of taking this suggestion seriously was the practice in the US under the Reagan Administration (1977–85) of requiring benefit–cost analyses of all new regulatory initiatives. The requirement of positive welfare consequences is also frequently evoked in other contexts, including in the decision-making in the courts and administrative agencies. Sometimes the requirement translates into formal benefit–cost analyses, and at other times into a more general balancing of the costs and benefits of suggested policies.

It often appears less controversial to suggest policy approaches that accomplish particular policy goals at the lowest cost, but also here we encounter problems. In practice, these suggestions translate into advocacy for pollution permit trading systems and other policy instruments based on market incentives. A parallel line of reasoning in the area of risk policy draws attention to the marginal costs of reducing different risks. This line of reasoning calls for addressing first those risks that are the least expensive to mitigate. It may also suggest that we should allocate risks, including environmental risks, to the parties that can avoid them with the lowest possible costs (Calabresi, 1970). However, the choice of a policy instrument to meet a given policy goal is not free of ethical problems. The choice frequently involves difficult issues related to the way different instruments structure the relationships between different groups of people, and distribute the costs and benefits of a policy between them. For example, the popular discontent with trading systems to a large degree relates to the way they distribute benefits and costs between the public and the polluters.

There are good reasons to be wary of suggestions to use welfare criteria as a guideline in policy and project choices. First, and perhaps most fundamentally, there is no reason why some notion of individual or social welfare should be considered decisive by all agents when policy choices are made. Sometimes we do – or arguably should be able to – regard the welfare of other humans (or nonhumans) as more important than our own welfare (or even some hypothetical notion of social welfare). At other times, even these other-regarding consequentialist motivations may not be decisive: we may want to behave in a way that preserves or heightens our self-respect, and we may have firm ideas about what it means to be honorable or virtuous. The advocacy of welfarism as an exclusive guide for policy choices suggests that we should simply ignore these other concerns and their advocates as irrational when making collective choices. Yet a number of individuals may have these concerns and rightly feel insulted when they are silenced on the basis of *a priori* judgments from economics. Several scholars have also warned that the conceptualization of collective choices as exercises in welfare maximization may tend to socialize individuals to act increasingly as welfare-maximizers and to advance further commodification of the realms that are important for human life and well-being – an unwelcome prospect in their eyes (Radin, 1996; Hodgson, 1997).

One consequence of the reality of plural motivations is that policy problems are viewed differently by different groups of people, and also *vis-à-vis* other policy problems. There may indeed be policy problems that are viewed from a welfarist viewpoint by a great majority of people. Welfarism may thus not be a controversial guide to policy choices related to these problems. Yet there are many other problems that are not so viewed. Risks provide perhaps the best illustration here. Individuals may knowingly and voluntarily accept certain risks, such as those associated with driving a car or mountaineering, while rejecting involuntary exposure to smaller risks related to occupational and environmental health. They also may have different attitudes toward old and new risks (Huber, 1983), usually feeling a need for a greater degree of protection against new risks. Thus the debate about mitigating the risks of driving or flying may be quite different from the debate about the use of genetically modified organisms. The economist's policy prescriptions may be contested precisely because they fail to acknowledge the different moral character of different policy problems.

Even when welfare concerns do dominate a policy debate, there is no reason why those adversely affected by the prospective policies should accept a welfare criterion, the potential Pareto improvement, as an unambiguous guide to policy choices. The strong focus on allocative efficiency in economic analysis translates into inattention to distributive consequences. This bias among economists is justified on the mistaken belief that efficiency is ethically neutral (Bromley, 1990). Ironically, public policy decisions are fundamentally – and often quite consciously – distributive choices, because public policies redefine the endowments and future economic agendas of different groups of people. When the winners of a new policy outcome do not actually compensate the losers, the latter have good reasons to reject both the compensation principle and the policy alternative suggested by it. Not surprisingly, it matters a great deal *whose* welfare is improved and *whose* is impaired. The nature of environmental problems as interest conflicts is obviously relevant also when motivations are plural.

Hypothetical valuation

The second contested practice of environmental economics – monetary valuation of the environment (or components of it) – arises from a related aspect of welfarism. Economists will usually suggest monetary valuation of the environment in order to make judgments about the welfare consequences of policies or projects that have some nonmonetized aspects. This is common not only for public policies and projects related to the environment, but also for other public policies and projects as well.

The practice of monetary valuation is older than the scholarship in environmental economics. Public health professionals and campaigners frequently invoked monetized benefits of lower mortality and morbidity early in the 20th century to promote new public health initiatives such as the purification of drinking water. Such estimates were also important in downplaying older programs intent on disinfecting the dwelling and belongings of deceased persons (see, e.g., Whipple, 1908). The first systematic method for deriving monetized nonmarket benefits – the travel cost method – was developed to measure the positive welfare effects from the provision of national parks and other recreational facilities (Knetsch and Davis, 1966). The practice of monetary valuation of the environment gained considerable momentum in the late 1980s. Today, methods of monetary valuation are used for a wide variety of valuation problems, ranging from the valuation of endangered species to the value of human life. The wider use of valuation methods has engendered ethical problems, such as those related to the transferability of results from one valuation context to another, and the adjustments needed to transfer results from a high-income context to a low-income context. For example, in one study the value of a statistical life of a Russian is "conservatively estimated" to equal one thirtieth of the value of the statistical life of an American (Larson et al., 1999).

To the extent that monetary valuation serves the making of welfare judg-ments, the criticisms of welfarism apply here as well. That is, there is no compel-ling reason why the size of monetized net benefits should be decisive when considering public policies and projects. Furthermore, even when monetary consequences are important as proxies for welfare changes, the incidence of bene-fits and costs may be far more important than the level of benefits and their relationship to costs. For example, finding that preserving a species of great apes generates a hypothetical welfare improvement is not terribly useful in bringing about that result if the benefits accrue primarily as psychological satisfaction to affluent, concerned, and well-informed inhabitants of the developed countries, while the costs are borne as reduced or lost economic opportunities by the rural inhabitants of the pertinent developing countries where the apes are found. Indeed, others will insist that we do not need the valuation exercise at all to accept the moral obligation to preserve these particular endangered species. In response to this, an economist might well insist that "choices" must be made – and that saving the great apes might mean that it becomes impossible to save the black rhinoceros. And in response to this a philosopher might well insist that this idea of a "tradeoff" is incoherent. Indeed, several authors in this volume address this very point.

Yet monetary valuation has also other problematic implications that are re-lated to how it understands individuals, what role it constructs for them, and how it positions them in relation to collective choices. Most valuation methods vest individuals with preexisting preferences and perfect knowledge, thereby suggesting that their willingness to pay for environmental benefits exists *a priori* and that the research challenge is to *uncover and harness* those values for making decisions about the environment. However, it stretches credulity to imagination that all of us have well-informed preconceptions about environmental values. We need to learn and develop preferences when we confront difficult, novel, and often unique environmental policy choices. Monetary valuation of the envir-onment simply ignores this in the quest to harness the individuals' preexisting preferences in the form of their willingness to pay for the purposes of making collective choices. Valuation surveys actually confront respondents with the task of learning and making up their minds about their willingness to pay for a hypo-thetical environmental benefit for the specific purpose of the valuation study. It is a good question how much the respondents learn during the valuation exercise about the choice they are confronted with, and whether the willingness to pay that they come up with is useful for making collective choice (Vatn and Bromley, 1994).

Moreover, valuation studies regard respondents as consumers who "shop around" and pay for environmental benefits as they would other goods and services. However, this role may actually insult respondents, because it denies them participation in collective choices in the usual sense – as voting citizens or as "political entrepreneurs." The respondents' own behavior in valuation studies seems to support this sort of reasoning. The frequency of protest bids and other behavior that is not compatible with the preconceptions of economics in valu-ation studies suggests that the respondents are not satisfied with the role they are given in many valuation studies (Vatn and Bromley, 1994). Respondents may well feel that the issue at stake in the valuation study properly belongs in the political realm, where it should be addressed according to the conventional political procedures. This has been the general argument of many philosophers (Radin, 1996; O'Neill, 1997; Sagoff, 1988).

Discounting the future

The third contested area of environmental economics relates to the practice in benefit–cost analysis of discounting future costs and benefits of policies and projects being considered for undertaking at the present time. The economic logic behind discounting is simple enough. Most of us would prefer to have a certain sum of money today rather than waiting, say, five years to receive the same amount. Simply put, the "value" to us of the sum that lies in the future is not equal to the same sum today; if we do not receive it for five years then its real value to us is much diminished. By extension, it seems obvious to discount future costs and benefits in order to determine their present net value when considering public policies and projects. After all, the costs and benefits of public projects and policies are usually incurred at different times.

Discounting practices vary, and quite elaborate procedures have been suggested, depending on the specifics of the assessment task at hand. For example, different discount rates have been suggested for projects and policies that have short- and long-span consequences. Furthermore, it has been suggested that discount rates should reflect the type of risks that are involved in the assessed policies or projects, and that risks should be considered apart from the discounting procedure. Finally, different discount rates have sometimes been proposed for different stretches of the assessment period (Bazelon and Smetters, 1999).

Estimates of the present values of costs and benefits are often contested by insisting that the discount rate applied in the assessment exercise was too high or too low. There are also often arguments, depending on the undertaking, that the discount rate should have been zero. An increase or a decrease in the discount rate may indeed change the conclusions from the assessment exercise. Therefore, those interested in fewer public projects and programs can realize their goal through the choice of a relatively high discount rate. Conversely, those interested in projects that yield benefits in the far future (say, addressing problems of global warming) will advocate a lower discount rate. In other words, we see that the controversy over discounting may often reduce to differences in opinions concerning the desirability of particular choices and future outcomes. Those who care about future environmental benefits will often criticize high discount rates when they are applied to policies or projects that would generate such beneficial outcomes. On the other hand, they would quite likely be unhappy with low discount rates for nuclear energy projects, where the costs of decommissioning such plants would occur far into the future.

The fundamental question concerns the ethical dimension of discounting – an area that remains contentious within economics and philosophy. The practice of discounting mobilizes welfarism that characterizes environmental economics in

the context of intertemporal choices. It also commensurates all policy consequences whether they are commensurable or not in the minds of the agents that
participate in or are influenced by the policy choices in question. This ethical
commitment has significant consequences in policy practice, and thus it is no
wonder that discounting is contested so often. We would need to reconsider the
way in which we analyze and do intertemporal collective choices if we recognized that collective choices are not informed exclusively by welfarism, and that
not all policy consequences are commensurable and measurable in monetary
terms. This is the conclusion of several contributions in this volume.

We will now turn to the discussion on the ethical dimensions of applying
economic analysis to environmental issues in the contributions of this volume.

An Outline of the Book

This chapter forms the first, introductory part of the volume. Part II, "Economics, Ethics, and Policy Choices," includes four chapters by philosophers – and
philosophically oriented economists – offering different views of the nature of
public policy and economic analysis. In chapter 2, "Are Choices Tradeoffs?,"
Alan Holland challenges the traditional economic notion that all choices involve
tradeoffs. He questions the model of choice contained in standard economic
analysis of environmental policy, and he suggests a way of understanding choice
as a deliberated and creative action that is shaped by social context and institutions. Chapter 3, by Bryan Norton, is entitled "The Ignorance Argument: What
Must We Know to be Fair to the Future?" Norton critically examines the reduction of intertemporal ethical questions to those of appropriate levels of savings
and consumption in each generation, in the face of alleged ignorance about the
preferences of future generations of individuals. Norton argues that intertemporal
questions properly concern what values we should cultivate in future generations. This view stands in stark contrast to the standard economic approach, in
which the problem is cast as one of preserving the opportunities for future
persons to obtain the same (or higher) levels of consumption as we are currently
enjoying. In chapter 4, titled "Benefit–Cost Considerations Should be Decisive
When There is Nothing More Important at Stake," Alan Randall both challenges
the standard economic understanding of environmental policy choices and
confronts the many critics of this view. Randall argues that both economic
and environmental concerns should matter. He suggests that economic concerns
should be seen as decisive when important environmental concerns are not
threatened. He also argues that important environmental concerns can be acknowledged by framing choices by constraints such as the Safe Minimum Standard.
The final chapter in this section, chapter 5 by Juha Hiedanpää and Daniel W.
Bromley, is entitled "Environmental Policy as a Process of Reasonable Valuing."
Drawing from the economics of John R. Commons, the authors propose that
choices over environmental policy are best understood as the product of a process
of "reasonable valuing." That is, public policy arises because of a nascent feeling
that the status quo does not lead to "reasonable" environmental outcomes. Why

else do we suppose that in democratic states there is pressure to introduce new policies? In the face of this dissatisfaction with the status quo there will arise policy alternatives that will seek to address these objections. The process of searching for a new consensus will entail a balancing of conflicting interests – reasonable valuing.

Part III concerns "Ethical Concerns and Policy Goals." Here, the chapters challenge the standard economic approach to environmental policy that locates the individual at the center of environmental choice, and that then insists that utility-maximization is the proper goal of environmental policy. In chapter 6, entitled "Rethinking the Choice and Performance of Environmental Policies," Jouni Paavola argues that we should respect the individuals' well-informed preferences also when they are based on other-regarding or nonwelfarist values. Paavola also outlines the consequences of admitting nonwelfarist motivations and positive transaction costs for economic analysis of environmental policies. Chapter 7, by Olof Johansson-Stenman, is entitled "What Should We Do with Inconsistent, Nonwelfaristic, and Undeveloped Preferences?" Here, he argues that individual preferences are best ignored as a guide to policy choice when they are not well-informed. He defends enlightened paternalism that would seek to maximize the agents' welfare as distinct from the satisfaction of their preferences or the maximization of their utility. In chapter 8, entitled "Awkward Choices: Economics and Nature Conservation," Nick Hanley and Jason Shogren critically review the criticisms of standard economic analysis of environmental policy. They argue that while the critics' concerns are legitimate, cost and benefit considerations are valuable in policy choices and add important information to our understanding of the tradeoffs that we face in making those choices.

Part IV moves on from the consideration of policy goals and is entitled "Ethical Dimensions of Policy Consequences." This part contains four contributions by economists examining the interplay among ethical concerns for policy outcomes, the actual policy choices, and economic analysis of those choices. Chapter 9, by Allan Schmid, is entitled "All Environmental Policy Instruments Require a Moral Choice as to Whose Interests Count." Schmid focuses on the distributive dimensions of policy outcomes and indicates how the policy choices that bring them about require moral judgments. He exemplifies this argument by analyzing environmental measures based on common law, statutory environmental law, and the use of public spending and revenues. In chapter 10, entitled "Efficient or Fair: Ethical Paradoxes in Environmental Policy," Arild Vatn examines the problematic nature of trying to separate the efficiency and equity considerations in choices concerning environmental problems that are characterized by novelty, time lags, complexity, and high transaction costs. He also shows how instrument choice is imbued with the same problem. Vatn concludes that one cannot separate efficiency and equity, and that attention must be given to both. Chapter 11, by Bhaskar Vira, is entitled "Trading with the Enemy? Examining North–South Perspectives in the Climate Change Debate." Vira argues that efficiency and equity considerations cannot be separated in the negotiations on the establishment of global trading system for greenhouse gases. Vira demonstrates that the initial allocation of rights to emissions determines the level of benefits from a trading system at the same time as it resolves the distributive question. In chapter 12, "Social Costs and Sustainability," Martin O'Connor outlines an alternative approach to the analysis of environmental policy that combines deliberated social processes and the use of natural and social scientific information.

Part V of the volume is entitled "Ethics in Action: Empirical Analyses" and it seeks to substantiate the role of ethical concerns in the analysis and choice of policies and projects influencing the environment. In chapter 13, entitled "Empirical Signs of Ethical Concern in Economic Valuation of the Environment," Clive Spash discusses how contingent valuation surveys can be used to improve our understanding of the real motivations that agents have for environmental protection instead of only soliciting willingness-to-pay estimates. Spash also presents findings on motivations based on contingent valuation surveys conducted in the Caribbean, and in the United Kingdom. Chapter 14, by Andreas Kontoleon and Timothy Swanson, is entitled "Motivating Existence Values: The Many and Varied Sources of the Stated WTP for Endangered Species." They argue that crude willingness-to-pay estimates may not be useful for policy-making, because agents may be willing to pay for different and sometimes incompatible policy benefits. They also present findings on how conventional welfarist values and other-regarding values motivate positive willingness to pay for the existence of rhinoceros in Africa, and of the giant panda in China. Chapter 15, by Nick Johnstone and his associates, is entitled "Environmental and Ethical Dimensions of the Provision of a Basic Need: Water and Sanitation Services in East Africa." Their contribution examines the implications of direct versus amenity uses of environmental resources for policy choices. They argue that direct use of environmental resources for subsistence use complicates policy choices and introduces important ethical problems. They illustrate these ethical dilemmas by analyzing the linkages between the quality of surface and ground water, the provision of water supply and sanitation services, and public health in East Africa.

The sixth and concluding part contains the chapter "Economics, Ethics, and Environmental Policy" by Daniel W. Bromley and Jouni Paavola. The concluding chapter draws together a number of the major threads from the contributions of the volume, in order to bridge the gap between environmental economics and ethics, and to present one vision of a broader understanding of making environmental policy choices.

Concluding Thoughts on the Ethical Content of Environmental Choice

Persistent concerns about the ethical underpinnings and implications of conventional environmental economics emerge from its recommendations on a range of environmental policies. We believe that these contested realms would benefit from enhanced conceptual work at the intersection of ethics and economics.

The conventional economic approach to environmental issues clearly suffers from an overly simplistic view of human behavior. The reigning reductionist program of attributing all choice to welfare-centered motivations is clearly unsatisfactory (Sen, 1995). By focusing on the self-absorbed utility-maximizer, and by ignoring other-regarding and nonwelfarist behavior, this approach fails to explain human behavior, and it offers contrived stories with respect to actual motivations for human action. Equally problematic are the assumptions of well-formed, stable, and preexisting preferences, limitless cognitive capacity, and full knowledge about the choices faced by individuals. These assumptions effectively remove – finesse – the problems of learning and arriving at meaningful judgments about human action. Several contributions in this volume address these problems. For example, the contributors indicate that – depending on the context of choice – there may be good reasons both to respect individual preferences instead of individual welfare, and to ignore individual preferences and to consider individual or social welfare. Yet the implications of a broader and more sophisticated view of human behavior remain to be explored, as do the full policy implications of such view.

Several other weaknesses in standard environmental economics result from its simplistic view of human behavior. For example, if – as suggested here – there is a broad ethical basis for human behavior, individuals will situate different choices or acts in different ethical realms. For example, certain environmental issues, such as the preservation of a particular species, may be framed as noneconomic moral questions. Individuals are well able to address these issues with the logic of the realm of moral commitments. They may also be able to apply the logic of the economic realm to moral issues – as when requested in a contingent valuation survey, for example – although they would not find it congenial. They may also be able – and actually be forced – to recast the issue into the economic realm if their livelihood is threatened. Yet this shifting of decision realms may not be taken as evidence of the fundamental correctness of economic calculation but, rather, as an example of the discontinuities in how we perceive and approach different choice situations. That is, individuals may refuse to apply self-centered calculation to certain choices, and consider other-regarding behavior as decisive until the personal consequences reach some threshold level that triggers a change in the way they frame the choice. Regard for others does not entail disregard for one's self, nor does self-regard entail disregard for others. Human action (choice) is a balancing across realms of reason.

A more realistic stance in environmental economics entails accepting a complex view of policy problems and policy options. There will always be new phenomena presenting us with choices of unknown or ambiguous consequences both in the epistemic and moral sense. Thus, the idea that there are universal criteria of desirability applicable to all policy problems and choices must be seen as a manifestation of a singular lack of imagination on the part of those who are committed to this flawed notion of reductionism.

References

Arrow, K. J., Cropper, M. L., Eads, G. C., Hahn, R. W., Lave, L. B., Noll, R. G., Portney, P. R., Russell, M., Schmalensee, R., Smith, V. K., and Stavins R. N. 1996: Is there a role for benefit–cost analysis in environmental, health, and safety regulation? *Science*, 272 (12 April), 221–2.

Bazelon, C. and Smetters, K. 1999: Discounting inside the Washington, DC beltway. *Journal of Economic Perspectives*, 13, 213–28.

Bromley, D. W. 1990: The ideology of efficiency: searching for a theory of policy analysis. *Journal of Environmental Economics and Management*, 19, 86–107.

Calabresi, G. 1970: *The Cost of Accidents: A Legal and Economic Analysis*. New Haven, CN: Yale University Press.

Hodgson, G. M. 1997: Economics, environmental policy, and the transcendence of utilitarianism. In J. Foster (ed.), *Valuing Nature? Economics, Ethics, and Environment*, London: Routledge, 48–63.

Huber, P. 1983: The old–new division in risk regulation. *Virginia Law Review*, 69, 1025–1107.

Knetsch, J. L. and Davis, R. K. 1966: Comparison of methods for recreation evaluation. In A. V. Kneese and S. C. Smith (eds.), *Water Research*. Baltimore, MD: Johns Hopkins University Press, for Resources for the Future, 125–42.

Larson, B. A., Avaliani, S., Golub, A., Rosen, S., Shaposhnikov, D., Strukova, E., Vincent, J. R., and Wolff, S. K. 1999: The economics of air pollution health risks in Russia: a case study of Volgograd. *World Development*, 27, 1803–19.

O'Neill, J. 1997: Value pluralism, incommensurability and institutions. In J. Foster (ed.), *Valuing Nature? Economics, Ethics, and Environment*. London: Routledge, 75–88.

Palmer, K., Oates, W. E., and Portney, P. R. 1995: Tightening environmental standards: the benefit–cost or the no-cost paradigm? *Journal of Economic Perspectives*, 9, 119–32.

Radin, M. J. 1996: *Contested Commodities*. Cambridge, MA: Harvard University Press.

Sagoff, M. 1988: *The Economy of the Earth: Philosophy, Law and the Environment*. Cambridge: Cambridge University Press.

Sen, A. K. 1995: Rationality and social choice. *American Economic Review*, 85 (March), 1–24.

Vatn, A. and Bromley, D. W. 1994: Choices without prices without apologies. *Journal of Environmental Economics and Management*, 262, 129–48.

Whipple, G. C. 1908: *Typhoid Fever: Its Causation, Transmission and Prevention*. New York: John Wiley.

PART II

Economics, Ethics, and Policy Choices

2

Are Choices Tradeoffs?

Alan Holland

One refrain repeatedly voiced in the context of environmental decision-making is this: *choices have to be made*. Almost as common is this gloss: *we have to make tradeoffs*. Often, the second claim is converted into something stronger still, namely: *only if our choices have the form of a tradeoff will they be rational choices*. The aim of this chapter is to challenge the view that a choice is essentially a tradeoff. The reason why it matters whether choices are tradeoffs is that the tradeoff view licences certain approaches to environmental policy and decision-making – in particular, benefit–cost analysis and associated methodologies such as contingent valuation. But if the tradeoff view is misguided, then these approaches will need to be reassessed. To achieve the aim it will be necessary to challenge the theory of human nature and human motivation that underlies the tradeoff view, and which, for convenience, will be referred to as the "belief/desire model." In the light of our critique of the belief/desire model, an alternative, "contextual," model of human motivation will be sketched, and some remarks made about the contrasting policy implications of the two models.

The notion of "tradeoff" is used here in a particular sense. Sometimes we use the term "tradeoff" merely to indicate a situation that involves a "sacrifice" – a situation in which something has to be "given up." So a person who claims that we have to make tradeoffs often simply means to point out that life involves sacrifices. That is not the sense that is at issue here. Rather, what is being contested is a theoretically charged notion best expressed as the view that *all choice is basically a form of exchange*. In this context, "tradeoff" refers to the idea that, in any choice between a range of options, there is always a dimension of value in terms of which the options that we face may be compared as more, less, or equally to be preferred. And choice is "rational" just insofar as it directs us to what is preferred. Conversely, the view is sometimes advanced that in the absence of a comparative "yardstick" of this kind, the choice will be arbitrary, or "unintelligible" as a choice (Regan, 1997: 144). At the back of such a view lies the "belief/desire model" of reasons for action.

The Belief/Desire Model

According to the belief/desire model of practical reasoning, in its crudest form, both beliefs and desires are required to motivate an agent to act. Beliefs without desires are inert; desires without beliefs are blind. Desires give the agent the motivational push to move, while beliefs are channels that guide the move to the right place. My desire for an apple moves me to search for one, while my beliefs guide the search to the fruit bowl.

Versions of this model underpin the axioms of neoclassical economics. They can also be found underlying more sophisticated psychological analyses of environmental decision-making, such as that presented by Stern and Dietz (1994). On the one hand, these authors acknowledge a creative element in human decision-making: they regard "attitudes" as constructed in the light of beliefs about how environmental policies and interventions will affect items of value. On the other hand, these values themselves – "social-altruistic," "biospheric," and "egoistic" – are treated as mere "orientations" that "take shape during the socialisation process and are fairly stable" (Stern and Dietz, 1994: 66–8). This account of the antecedents of action contains strong echoes of the work of David Hume, who writes that:

> reason in a strict and philosophical sense can have an influence on our conduct only after two ways: either when it excites a passion by informing us of the existence of something which is a proper object of it; or when it discovers the connexion of causes and effects, so as to afford us means of exerting any passion. (Hume, 1978 [1740]: 3.3.1)

The general outline of the model was laid down by Aristotle, who depicted reasoned action as conforming to the pattern of what he called a "practical syllogism." Such syllogisms comprise (i) a major premise identifying some state of affairs as desirable (for example, "everything sweet must be tasted") and (ii) a minor premise or premises indicating the opportunity to bring the desirable state of affairs about (for example, "this [particular thing in front of me] is sweet"), yielding (iii) the enacted conclusion: "I'll have some" (Aristotle, 1999: 1147a30–2).

There is no denying the power of this model, nor its distinguished historical ancestry. Indeed, many seem to regard it as almost too obvious to require defence. However, its implications for public policy are quite profound, and far from innocent. For in the context of public policy, where the crudest versions of the model seem to hold sway, all choice is taken to be a kind of transaction, or exchange of goods. Human agents are judged rational insofar as they seek to maximize their utility, which is often construed in terms of preference-satisfaction. This project, in turn, is only possible if all options can be compared according to some common measure of value. The beauty of preference-satisfaction as the touchstone of utility is that it appears to afford a universally applicable measure of value – namely, the degree of preference attaching to each available option. Decision-making, whether individual or collective, is then a matter of collecting, aggregating, and weighing preferences. Ideally, therefore, according to the model, public policy should be determined by the extent to which it works toward the satisfaction of the most – or the most urgent – preferences.

In the light of such a model, benefit–cost analysis naturally becomes the instrument of choice for policy and decision-makers. And despite the acknowledged shortcomings of associated methodologies such as contingent valuation, it also explains why these shortcomings continue to be treated as technical problems only: in principle, it is assumed, the approach *has* to be sound. On the basis of this model, it is also but a short step to the view that the institutions that are most suited to deliver the aims of public policy are those of the market which, after all, are built upon the relations of transaction and exchange. Hence, Elizabeth Anderson is led to exclaim that "The dominant models of human motivation, rational choice and value . . . seem tailor-made to represent the norms of the market as universally appropriate for nearly all human interaction" (1993: xi–xii).

In legal and social contexts also, the influence of the model is pervasive, especially in generating the view that benefit–cost analysis is the best, and even perhaps the only rational method of approaching areas of potential social conflict. In recent UK legislation governing scientific experiments on animals, for example (Animals – Scientific Procedures – Act 1986: sec. 5.4), it is laid down as a statutory duty of the Secretary of State to "weigh the likely adverse effects on the animals concerned against the benefit likely to accrue from the programme specified in the licence." According to the Environment Act 1995, the newly constituted Environment Agency is likewise enjoined, as part of its statutory duty, to take into account the benefits and costs of any project or policy. It is hardly an exaggeration to say, therefore, that the "exchange" model of choice and decision-making is thoroughly institutionalized. It will now be contended that, however influential it may be, the belief/desire model of what it is to act for a reason is deeply flawed. And while there might well be sophisticated versions of the model that escape such criticism, the cruder forms that underlie many of our existing decision-making institutions do not.

A Critique of the "Belief/Desire" Model

In what follows, five grounds of criticism of the "belief/desire" model will be advanced. First, acceptance of the model carries with it some unwelcome implications. Second, desires, as construed in the model, fail to provide reasons when they should, and purport to provide reasons when they should not. Third, the model fails to make sense of practical reasoning. Fourth, the model begs the question against the possibility of incommensurable values. Fifth, and finally, what purports to be a model of practical reasoning is incapable of explaining either choice or action. Each criticism will be discussed below in greater detail.

Unwelcome implications

A first criticism of the belief/desire model is that it appears incapable of explaining certain common phenomena such as weakness of will and self-sacrifice.

There is a formal similarity between these phenomena. On the one hand, we can describe weakness of will as someone's (apparently) sacrificing his or her own best interests, where this is perceived as an *ignoble* thing to do; and on the other hand we can describe self-sacrifice as someone's (apparently) sacrificing his or her own best interests, where this is perceived as a *noble* thing to do. In both cases, the problem for the belief/desire model of human action is this: How can a person, knowing what it is best for him or her to do, willingly and knowingly not do it? And the belief/desire model appears to have no answer. It must say one of the following:

(i)   that the action chosen was "really" (perceived by the agent as) for the best,
(ii) that the agent was under some physical or psychological compulsion, or
(iii) that the agent did not know what he or she was doing.

A moment's reflection is enough to reveal that on none of these accounts is it possible for an agent freely and deliberately to sacrifice his or her own best interests, whether nobly or ignobly. A fourth option is to say that such an agent is (or has to be) irrational: which is to say that no account can be given of why he or she has acted in this way.

It must be conceded that the inability of the model to account for such phenomena is not a conclusive objection to the model – partly because the phenomena are inherently difficult to account for anyway, and partly because it is open to defenders of the model to question whether these are real phenomena. Socrates was the first, but not the only, philosopher to deny that weakness of will exists, and to insist that it must be due to ignorance on the part of the agent about what it is best to do. Not surprisingly, given his attachment to the practical syllogism, Aristotle's perplexing and inconclusive discussion of "*akrasia*" (weakness of will) comes close to reaching a similar conclusion (see Wiggins, 1980). However, in the case of both philosophers, it is possible to trace their position to some otherwise questionable assumption. Thus, Socrates' well-known tendency to equate virtue with skill arguably led to his placing an undue emphasis upon knowledge in his account of the virtues, and hence to construing any kind of vice – implausibly – as a form of ignorance. Aristotle's position, on the other hand, might be traced to a fundamental tension in his thought between monistic and pluralistic conceptions of the good life that was perhaps never resolved. A clear recognition of more than one, and possibly irreducible, ends would arguably have made it easier for him to assimilate weakness of will into his theory of practical reasoning.

As for self-sacrifice, cynics are quick to question the phenomenon: even Mother Teresa was "really" a selfish person pursuing her own self-interest. Without claiming to be able to resolve the issue here, we can at least defuse two arguments commonly advanced in defense of such a position. One is the notion that the pursuit of self-interest is somehow inevitable in a creature that is a product of natural selection. However, both the theory of kin selection, that demonstrates the possibility of a hereditary mechanism for altruism, and recent work by Elliott Sober (1995) that demonstrates the lack of any necessary connection between evolutionary altruism/egoism and psychological altruism/egoism, suggest that such a notion is no longer tenable. A simpler argument rests on the belief that selflessness has to be contrary to an agent's wants, together with a conviction that wanting is a necessary condition of voluntary action. This seems to be a plain mistake. What makes truly selfless people so admirable is precisely the fact that they do what they do *willingly*. Doing what you want (in this sense) and acting selflessly are not incompatible.

The point is simply this: there are no obviously strong reasons from outside the reach of the belief/desire model of human action for questioning the reality of these phenomena. So its inability to account for them is at least *prima facie* evidence for questioning the model.

There are also cases when it seems simply inept to construe choices as tradeoffs. When, for example politicians, or bank officials, are caught off-guard and speak of unemployment as a "price worth paying" – in other words, adhering faithfully to the interpersonal application of the tradeoff model – they are rightly castigated for showing a failure of sensitivity. They have misrepresented the character of the choice involved. Or consider again the choice offered to us by the highwayman who declares "your money or your life." (Clearly, these highwaymen are schooled in economic theory and have signed up to the tradeoff model.) Suppose a person refuses, and is shot. Is it really plausible to say that she values her money above her life? Not her money, you will hasten to say, but her dignity or some such thing. Even so, the idea that this is a form of exchange does not come easily. Rather, giving way in a situation such as this is simply unthinkable, not compatible with the woman's perception of the kind of person she is. So she "sacrifices" her life. And sacrifice is precisely not trade: it involves giving up (what the belief/desire model would hold as) "the more valuable." As Steven Lukes observes: "Trade-off suggests that we compute the value of the alternative goods on whatever scale is at hand, whether cardinal or ordinal, precise or rough and ready. Sacrifice suggests precisely that we abstain from doing so. Devotion to the one exacts an uncalculated loss of the other" (1997: 188). For the defender of belief/desire, there remains the defense that such actions are really irrational and therefore unintelligible – a possible defense, yes; but also a rather desperate one. Can we so lightly set aside some of the most noble and inspiring actions of which humans are capable?

Reasons and desires

A second problem stems from a central feature of the belief/desire model – namely, that it separates the cognitive and noncognitive components of human motivation. The problem is that, shorn of any cognitive content, desires become indistinguishable from brute urges, with the result that they are unable to constitute reasons for action at all. Warren Quinn imagines someone who has a strange functional state – an urge – that disposes him or her to turn on any radio he or she sees to be turned off (1995: 189). Such a person, he rightly claims, does not even have a *prima facie* reason to turn on a radio: nor, one might add, does a bystander have any reason to aid and abet this person. There is a striking anticipation of the point in the rich analysis of human motivation to be found in Plato's *Republic* (see Plato, 1993). In Plato's view, although desires can constitute reasons for action, they do so only in the context of a well-ordered and hierarchically structured psyche, where considerations of self-respect and some overall conception of the good moderate the extent to which, and the manner in which, they find expression. Desires that come seriously adrift from that context, as happens in the process of psychic disintegration (*Republic*, Books 8 and 9), become insatiable and take on the form of an addiction. In that form they explain action, but can no longer justify or provide reasons.

An associated problem is that the belief/desire model makes no differentiation among desires: making due allowance for strength, every desire constitutes a *prima facie* reason for action. But this is implausible, as Quinn indicates with his example of the pyromaniac who has a passion for setting fires, but at the same time hates himself for taking pleasure in such activity (1995: 191). The tradeoff model simply fails to explain the power of one desire to moderate, and even wholly to negate, the justifying power of another. According to the model, each desire constitutes a reason, and the agent has reason to satisfy whichever is the stronger. But while it may be true, in Quinn's example, that the presence of the "higher-order" desire fails to extinguish the pyromaniac's desire to set fires, what it surely does do is extinguish its capacity to provide a reason for action. Hence desires, as construed by the model, not only fail to provide reasons when they should, but purport to provide reasons when they should not. Plato's equivalent case involves a certain Leontius, who is disgusted at himself for ogling the corpses of newly executed criminals (*Republic*, 439e). He uses the case to argue for his hierarchical model of human motivation: desires lose their capacity to afford reasons when they are not moderated by other sources of motivation. Plato also shows how the gratification of certain desires – for example, the securing of money by means of some shameful act – can operate causally to undermine the capacity of the agent to sustain these other sources of motivation. Hence the capacity of the so-called "decent" desires to provide reasons for action is corroded. It ought to be disturbing to proponents of the tradeoff view that, according to Plato's analysis, the absence of "trade barriers" between all the various springs of human action – the view that is implied by construing choices as tradeoffs – is the mark of a psyche that is moving toward total disintegration.

Making sense of practical reasoning

A further criticism of the model is that it treats desires as "preformed" and "given" prior to the exercise of choice, rather than as "formed by" (or open to formation by) the exercise of choice – a position that scarcely withstands scrutiny. Basically, it makes the origins of our desires mysterious, and makes it difficult to hold ourselves responsible for our actions. Aristotle remarks that "we become just by doing just acts". As Miles Burnyeat (1980) explains in his persuasive analysis of Aristotle's account of how we come to be good (or bad), the thought behind Aristotle's remark is that we learn the delights of good behavior through doing the right thing on particular occasions: in other words, we develop the desire to do the right thing (and equally, of course, the wrong thing) through practice. In this way, our desires do not simply assail us; rather, we carry some responsibility for their development and therefore for the character that they serve to form.

This is but part of a broader problem, which is the difficulty that the belief/desire model has in making sense of the discursive antecedents of choice and action – in a word, with practical reasoning as such. Perhaps the chief casualty is the process that lies at the heart of practical reasoning – namely, deliberation. The belief/desire model generates the notion that actions result from, and are preceded by, discovering our most powerful leanings – and therefore that deliberation is a process of discovery and report. But as Joseph Raz shows, there are at least two things wrong with this approach. One is that it casts our desires as somehow out of our control. The other is that it makes no sense of deliberation, which is not a process of discovering what we want, but a process of reflecting upon what there is most reason to want (Raz, 1997: 115): to construe deliberation as discovery is to misrepresent its logic. There is collateral damage too to the judgments that issue from deliberation, and inform both choice and action. It has been argued (Holland, 1997: 488, 491) that these too are open to critical assessment in the light of a range of criteria – of relevance, attention, sensitivity, and so forth; also that we may be called to account for these judgments and must take responsibility for them. It is hard to see, then, how what we might call the "texture" of practical reasoning, and especially the deliberation and judgment that forms its core, gains adequate expression through the belief/desire model. Rather, as David McNaughton puts it in his account, the aim of practical reasoning is "to organise . . . competing conceptions of a situation into an overall picture in which the various considerations find their proper place" (1988: 130).

Desires and preferences do not arrive with some preestablished "weight," lending themselves to comparative assessment. Nor should the resolution of conflicts between desires be structured along mechanical lines, as some kind of "resultant of forces." Desires function as reasons *only if* there are reasons for the desires. Accordingly, the resolution of conflicts between desires, whether within or between individuals, requires a "reason-sensitive" process rather than a mechanism of exchange: the point is not how strong or widely held the preference is, but how strong and relevant is the reason for the preference.

Incommensurable values

A crucial challenge to the belief/desire model is posed by the fact that many choices involve values, together with the possibility that many values – that is, judgments about what it is good to do or to be – are "incommensurable". By "incommensurability" we refer to an intelligible choice between feasible options, where there is no appropriate value in terms of which the options might be compared as "better," worse," or even approximately equal. As Raz puts it, two goods are incommensurable with respect to some scale if one is neither better, worse, nor equal in value to the other in the respects measured by the scale

(1986: 322). Notice that the concept of incommensurability, so defined, is more radical in its implications than the increasingly familiar notion of lexically ordered choice. Lexical choice refers to a situation in which the decision space is "partitioned." On the one side are considerations that will be addressed when, and only when, all the considerations on the other side have been fully addressed. If that is how our desires and/or values are structured – and there may be more than one "partition" – then scope for tradeoff will be limited accordingly. But still the "partitions" are ranked, and one or other will have priority; there can also be tradeoffs within the partitions. With incommensurability, on the other hand, there just is no account available of the relative standing of the values concerned. The dilemma discussed by Jean-Paul Sartre – that of the young man who has to decide whether to join the resistance or stay behind and care for his mother – is a good example. He has to choose, it appears, between love and duty, two values between which, on the face of it, there appears to be no commensurating value, and pending further description, no obvious order of priority.

What is true, at any rate, is that the assumption of comparability can no longer be taken for granted. In Ruth Chang's anthology, *Incommensurability, Incomparability, and Practical Reason* (1997), the majority of philosophers are all lined up *against* comparability. The beginning of Donald Regan's contribution to the volume says it all: "In this volume I am the 'designated eccentric,' appointed to take a position no one else would touch with a barge pole. I believe in what I shall term 'the complete comparability of value'" (1997: 129).

Faced with incommensurable values, some would argue that rational choice is impossible. It would seem impossible, at any rate, by the lights of the belief/desire model of practical reasoning. The existence of incommensurables is therefore held to pose a threat to rational decision-making. This, at any rate, is the view of Ruth Chang (1997: 13). Her position depends upon a certain view of what justified choice entails, which is recognizable as a version of the belief/desire model. In her version, a justified choice either is, or depends on, "a comparison of the alternatives with respect to an appropriate value." This is the view that would make all choices that entail an element of conflict into tradeoffs. If all choices involve "a comparison of the alternatives with respect to an appropriate value" then, in cases of conflict, the "appropriate value" in question provides the unit of exchange – the respect in which the course of action chosen leaves one "better off." It is Chang's attachment to this model of practical justification that leads her to deny the reality of incommensurable values: she regards it as essential if there is to be such a thing as rational choice.

Donald Regan also takes this view. According to Regan, in the absence of some comparative yardstick, the choice between A and B will be arbitrary, or "unintelligible" as a choice (Regan, 1997: 144). There is therefore no conceptual space for the phenomenon that we recently tried to describe – an intelligible choice between incommensurable values: there has to be a comparative judgment that there is *more* reason to do A than B – or at least no more reason to do B. Otherwise, Regan claims, the agent "will have no way of making what happened fully intelligible to herself as her choice." The reason for this is that "choice between two specific goods must be based on reason to prefer one of the goods to the other." Otherwise, it is not a choice at all, but simply something that "happened to the agent" (ibid.).

Now, if incommensurable values really were a threat to rational decision-making, that would indeed be reason to question their existence. But what they actually threaten is the belief/desire model. And it would seem equally plausible, if not more so, to turn the argument on its head. Rather than assume the model and deny incomparability, it may be proposed instead that we take the idea of incommensurable values seriously, and take equally seriously the threat that this poses to the belief/desire model of choice. If there are such values, the model falls. But even the possibility of such values discredits the model, since there is no obvious contradiction in supposing them to exist, and the model is committed in advance to finding decisions in which they figure unintelligible. The question we should be asking is: What model of practical reasoning, if any, will enable us to make sense of choices that involve incommensurable values? Indeed, such a test could be proposed as a criterion of adequacy for any theory of human agency.

It is perhaps worth insisting at this point that to challenge the view that choices essentially involve a tradeoff or exchange is in no way to deny the existence of tough decisions – quite the contrary. Part of the point of the challenge derives precisely from the fact that the exchange or tradeoff model *fails utterly to explain the toughness of tough decisions*. Indeed, such models serve only to conceal and suppress the toughness of choice. The point can best be registered by considering the typical "fallout" from a tough decision – namely, anguish. Suppose that tough decisions are indeed tradeoffs. As previously mentioned, this certainly means that something perceived as desirable has been given up or foregone. And no doubt this gives us cause to regret what we have had to give up. But at any rate the exchange has been made, and we have got the best deal. There are hardly grounds here for anguish over the decision itself. Yet anguish is precisely what one might expect in the wake of a truly tough decision. This may stem in part from what Bernard Williams has called the "residue" of tough decisions – perhaps a perception that whatever we do would be wrong – but it also stems in part precisely from the *absence* of a yardstick, a circumstance that leaves us lost and confused, can induce trauma, and can even break our spirit.

The problem of explanation

A final criticism focuses on the question of how far one *explains* a choice or an action by reference to a desire or to a tradeoff. Consider, first, how we might go about explaining the most banal of choices – say, the choice between vanilla and raspberry ice-cream.

Q.   "Why did you decide to have the raspberry/vanilla?"
A1. "Because I prefer the raspberry to the vanilla/the vanilla to the raspberry."

One thing that immediately strikes one is the *emptiness* of this as an explanation.

A2.   "Because I prefer the taste of vanilla/raspberry."

Now, perhaps, we are beginning to get somewhere: the explanation has something to do with the taste – rather than, say, the colour.

A3.   "Because I prefer the full-bodied/astringent taste."

Let us suppose that fullness of body would justify the choice of vanilla, and astringency the choice of raspberry. I wish to make two claims:

1   It is only when we get down to the reference to fullness of body or astringency that we are really beginning to *understand* what happened and to *understand* the grounds for the choice.
2   Fullness of body and astringency, although they are both qualities of taste, are not (obviously) comparable. They are *different* qualities of taste, but there is no quality of taste such that one possesses more of it than the other.

The inference to be drawn is this: if even this most banal of choices does not lend itself to being explained – that is, made intelligible – in terms of a preference or tradeoff, then it seems unlikely that choices in general can be explained in this way. This point leads to what is perhaps the central claim of the chapter: that to construe choices as tradeoffs is to construe them in a way that is empty of explanatory significance.

The only sense in which reference to a desire or preference can explain an action, I suggest, will be a sense that *contrasts* with a case in which we intentionally do something that we would prefer not to do. In other words, it must qualify the intention in some way. Otherwise, reference to a bare desire explains nothing. It serves merely to signal *that* an action is intentional; it does not signal *what* the intention is. The idea that reference to a desire, or to a preference, or to a greater "weight" or "importance," is necessary to explain an action stems from the assumption that, to justify or explain an action, one needs, so to speak, to kick-start what is inert. But if that really were the problem, it is not clear that anything on earth could provide the solution. The point is captured in Thomas Scanlon's discussion (1998: 20–2) but can be traced back to Aristotle's identification of the category of "voluntary," as distinct from "involuntary," action. It is only to actions *already conceived of as voluntary* that it makes sense to assign a reason. Choice, for Aristotle ("*prohairesis*" – the decision consequent upon deliberation) is a subclass of the voluntary. A reason could neither justify nor motivate something that was a brute movement. As Scanlon observes: "Actions are the kind of things for which normative reasons can be given only insofar as they are intentional, that is, are the expression of judgement-sensitive attitudes" (1998: 21).

Thus, if someone were to ask you "Why are you sitting in that armchair reading Dickens' *Pickwick Papers*?", and you were to reply "Because I want to," then far from having given your reason for sitting in the armchair, you would, I suggest, be indicating *your unwillingness to give a reason*. Far from being a paradigm of what it is to give a reason for one's actions, to say "Because I want to X" is to give no reason at all. In asking for a reason, the questioner is already taking it for granted that there is intentional agency, since this much is presupposed in the quest for a reason. The point of the question is to elicit some further specification: what it is about what the agent is doing that is construed as desirable, or as a good reason for the action. It is this, not a statement of preference, that will make the action intelligible. Reasons, then, are not open to quantification: there is no function such that having more reason to X is necessarily correlated with the presence of some valued feature to a greater degree. Incommensurability is not the exception; it is the rule.

The prevalence of utilitarian modes of thought may explain why the opposite view is so common. One of the advantages frequently claimed in elementary accounts of that theory is that it provides a universally applicable "decision procedure." Happiness, or freedom from pain, is presented as the common measure, in terms of which all options can be evaluated. However, any serious attempt to apply the theory in practice will reveal the deception. Happiness is not a homogeneous item, but a mosaic of heterogeneous elements. There just is no common substance – no _utility_ – by which to compare, for example, the suffering experienced by an experimental animal with the understanding gained from the experiment. Nor is this a point about moral reasons only, but about reasons generally. The determined egoist, confronting a chocolate bar that will ruin his or her waistline, will soon find that he or she has to decide between vanity and greed, and will just as surely fail to find an "appropriate" value in terms of which to compare the alternatives. Self-interest is not such a value, since it is as heterogeneous an objective as happiness. There is a close analogy here with (Darwinian) "fitness." The temptation to think that a comparison of value explains action echoes the temptation to think that fitness explains survival. In both cases, the explanatory work is done by more specific descriptions: astringency explains the choice of raspberry; length of neck, or the precise shade of blue on the rump, explains survival and reproduction. The ranking of value, and the fitness, are descriptions applied only _after_ the event.

As a corollary, the very idea that rationality is built in to the notion of a tradeoff becomes suspect. The model invites us to frame the rationale for all decisions, in terms of a "majorizing" or "maximizing" strategy – the idea that an agent must always be construed, if his or her actions are to be intelligible, in a quest to increase or maximize. One robust response is to contend that any majorizing or maximizing strategy is in itself inherently _ir_rational. The reason is simple. The policy of pursuing more, or the most, of something, literally has no "ratio" – there is no answer to the question "How much more of something will be enough?" But if there is no answer to the question "How much more of something is enough?" then how can it be a reason for doing something that one is providing "more"? The results of ignoring such considerations are evident in the problems that beset much health provision in modern Western democracies. There is no longer enough health care to go round, and the "solution" usually takes the form of some kind of rationing: an irrational solution to an insatiable (literally = "never having enough") and therefore irrational demand.

What economists know as "diminishing marginal utility" is not a "brute" phenomenon, but an empirical symptom of the emptiness of the quest to increase or maximize for its own sake.

Features of a More Adequate Model of Human Agency

In attempting to delineate the features of a more adequate model of human agency, the task is not to correct a "mistaken picture," or to replace a "false" picture with a "true" one. That would be to assume a fixity about human nature that is quite at odds with the stance being proposed here. We shall assume, to the contrary, that as reflective and self-conscious beings, we have it in our power to modify our natures in a variety of ways, and this includes the ability to deceive and to delude ourselves. We can play the part of victims, and we can indeed play the part of "rational" economic agents. But this would be a kind of choice, even an ideological choice – a matter of *deciding upon* the direction in which to take our human natures.

In light of the deficiencies identified in our discussion of the belief/desire model, we can suggest that a more adequate model of practical reason and agency will construe it as follows:

1 A reason-sensitive and social phenomenon that is responsive to social and institutional context.
2 An engagement in practices whose ends are internally related to the practice, and that are therefore (often) aimed at what is right, fitting, sufficient, or appropriate for its own sake.
3 Motivated by self-respect, and keyed into notions of identity and integrity.
4 An experimental and formative phenomenon that is therefore expressive of curiosity, creative, and interested in self-discovery.
5 A process of maneuvering and of renegotiation of available choices, so that a measure of independence and autonomy can be affirmed.
6 Part of a continuing process of interpretation and reinterpretation that seeks, among other things, to make, and give, sense to the past – both of the agents involved and of the institutions to which they belong.

1 Human agents are sensitive both to reasons and to context, and will modify their behavior accordingly. In view of this fact, their behavior is unlikely to be accounted for satisfactorily in terms of (i) fixed goals or drives/orientations (for example, egoistic, altruistic) and (ii) opportunities for their realization. Context, in turn, might be understood in either social or institutional terms. From a social point of view, relevant considerations will include the behavior and habits of others, and relationships – of power, affection, and so forth. Thus, in a household of untidy people, a given individual may not bother to be tidy. In a different context, they would. Since actions are rarely isolated, it is not surprising if the actions of others affect the very meaning of a given individual's actions. Think, for example, of how the actions of others will make the difference between "a

hopeless gesture" and a "demonstration of solidarity": context changes the *meaning* of what is done. Institutional arrangements are also crucial – particularly because of the values that inform them: people may be more willing to cooperate, for example, under a system that they perceive as fair. For this reason, asking individuals, in isolation from any social context, how they will behave (as in contingent valuation studies, asking what they would be willing to pay) is unreal. In more obvious vein, ongoing external constraints are of particular importance when these take the form of statutes, regulations, and the like. And little, if any, decision-making takes place except within the context of such constraints.

These informal considerations gain credence when placed alongside more painstaking empirical studies. Since R. M. Titmuss (1970) first suggested that monetary payment might undermine people's willingness to give blood, there has been growing empirical evidence for what has come to be called "crowding out" – the tendency for monetary inducements to undermine the intrinsic motivation that people have to act in cooperative or public-spirited ways (see especially the work of Bruno Frey, 1994, 1997). A recent example comes from Israel, where parents were tending to arrive late to pick up their children from a day-care centre (Gneezy and Rustichini, 2000). After a substantial fine was imposed for late collection, in the expectation that this would deter parents from arriving late – as standard economic theory might predict – there was found to be a significant *increase* in late collection, contrary to expectation. In fact, the parents now saw themselves as "entitled" to arrive late, since they were paying for the privilege. The institution of the fine had changed the nature of the relationship: the social and institutional context was now different.

2   Agents will do what is right just because it is the right or decent thing to do, and for no other reason. This is the intrinsic motivation referred to above. Although it has become fashionable in some quarters to question the existence of such motivation – and Kant's rather severe talk of duty for duty's sake may have contributed to the disenchantment – it is worth reminding ourselves that Aristotle, for one, thought it a crucial component of the exercise of virtue. In the absence of such motivation, there would be no such thing as virtue. So, for example, a courageous act done for the sake of gain does not count as the exercise of virtue, but only if it is done *because* it is the courageous thing to do. Such acts will bring their own rewards, but it is not for the sake of the rewards that they are done (and if they were, the rewards would not accrue). The existence of such motivation poses a clear threat to the tradeoff model: the rightness of an action remains a sufficient reason to do it, whatever reasons there are for doing otherwise. In similar vein, agents will also seek to do or bring about what is fitting, sufficient, or appropriate, because it is fitting, sufficient, or appropriate. This is a point about the character of the object of desire, and if true will again threaten the tradeoff model of choice. For if a person settles for what is sufficient when something of greater value is on offer, he or she must appear, from the point of view of the tradeoff model, to be behaving irrationally. For the theoretical basis of this position, we need look no further than Aristotle, whose doctrine of the "mean" – his attempt to characterize the exercise of practical wisdom – captures

this aspect of decision-making precisely. Although Aristotle is well known for his view that all human action is aimed at the "good," he never thinks of this as a maximizing pursuit. Happiness ("*eudaimonia*") is described as the ultimate pursuit because it is not sought for any further reason, and is "self-sufficient." Enough *is* enough. His model is, rather, that of "hitting a target." Anyone who is exercising practical wisdom is acting for the best of reasons. And this, says Aristotle, is a matter of doing the right thing, at the right time, in the right way, and in relation to the right thing: this is reason enough.

3   It is a mistake to think of decisions and choices as only *issuing from* agents or groups of agents. Rather, they help to define and create notions of identity and integrity, especially if they are the means to retaining self-respect. These facts simply cannot be represented in terms of exchange and tradeoff. In the language of William James (1904: 30), who was inspired by the electrical model of "live" and "dead" wires, some "possible" beliefs and some "possible" courses of action are simply not "live" options. In the case of beliefs, it is because they are unthinkable. In the case of actions, it is because they are intolerable or unendurable: they are not compatible with personal integrity and the retaining of self-respect. Reasons here are categorical, not comparative. To opt for A (divorce, migration, death) rather than life cannot be explained as a comparative judgment – a preference for A; it is explained by the fact that the alternative is intolerable. Once more, our theoretical guide is Plato, who contends that it is only a person who is no longer governed by his or her rational element and no longer retains his or her self-respect who will live life as if it were simply a matter of trading the more for the less advantageous option. True, there are magnificent bargains to be had: witness the case of Faust's bargain with Mephistopheles. But sooner or later, as Plato tells the tale, such a person will, for example, be ready to demean him- or herself for the sake of money. Self-respect acts as a source of motivation that moderates other motives; it cannot enter the market alongside these other motives without a risk of untold structural damage.

4   To think of action only in terms of the instrumental means to achieve some goal is a woefully partial conception. Action may, or may also be (for these are not exclusive roles) experimental, creative, and expressive of curiosity. Witness the sentiment of the early twentieth century economist Frank Knight, who declared that man is "an aspiring rather than a desiring being," who seeks "not satisfaction for the wants which he has, but more and 'better' wants" (Knight, 1922). We often feel that the question "Did I do the right thing?" or "Was it a mistake?" is inappropriate, and can more helpfully be replaced by the question "Can I/we make it work?" And it makes good sense, sometimes, to respond to the question "Why did you do that?" with "To see if I could." These features of motivation display the limits of preference-satisfaction as an objective, besides making a nonsense of aggregation and tradeoff.

5   Practical reasoning is also creative in another way – as a process of maneuvering and of renegotiation of available choices so that a measure of independence and autonomy can be affirmed. It is a matter of acting in such a way as to retain a sense of agency – surely a precondition for people to feel responsibility for, and

engagement with, both the process and the outcome of decision-making. What is at issue here turns on two radically different conceptions of "negotiation," both of which have particular application to the case of "tough choices." The negotiation proposed by the belief/desire model has the character of a bargaining process – striking the best deal. But negotiation as conceived on the alternative model refers to something quite different – a process of "finding one's way through." It is the sense we have in mind when we talk of negotiating "white water" in a canoe. Perhaps the situation can be recast to avoid the choice altogether. (We might drag the canoe overland for a short stretch.) People have learned to be wary, for example, of the choice that is sometimes presented between jobs and wildlife, and are suspicious of the ideology that informs such a "choice." Failing that, there are different pathways through. Here it is not, or not only, a matter of minimizing losses and residues, but a matter of doing justice to the various claims, where again considerations of appropriateness are to the fore; it is also a matter, simply, of "staying afloat."

6   Finally, it is well to recognize that decision-making, individual or collective, is a continuing process rather than an event, sometimes but not always, punctuated by "choices" and "decisions." Even choices and decisions themselves are probably not best thought of as events: thus, the question as to *when* a decision was taken might well not have a sensible answer. Against this background, we can see that many of the objectives of decision-making are actually internal to the process itself, although sometimes these are hidden from view. The (undeclared) purpose of some decisions may be to make sense of previous decisions, and sometimes also to vindicate those earlier decisions. The purpose of some decision-making procedures is actually to defer, or even entirely to avoid, reaching a decision; this may become explicit when a decision-making body, after deliberation, recommends a moratorium (cf., the US President's National Advisory Bioethics Commission in the case of human cloning). In other cases, they may be designed either to display attention to the matter in hand or to distract attention from more embarrassing matters. Decision-making needs to be seen against a broader background in which agents are seeking to interpret and reinterpret events, and to make sense of, and give sense to, their own pasts and those of the institutions of which they are part.

The Institutional Implications

The prevailing behavioral model construes the human subject as a utility-maximizer whose behavior is driven by preferences that are viewed as both determinate and "given," and that enjoy a certain authority and immunity from critical attention. Both individual and public deliberation is construed as a calculative exercise practised by individuals, in relative isolation from one another, who are engaged in transactional processes – for example, of negotiation and bargaining. Exchange is effected through the common currency of preferences. The epistemological input is conceived in the form of "information" that is not so much objective as impersonal and unattributed – the source is supposed to be unimportant. Equally unexamined are the existing assignments of rights and distributions of income that provide the context for any particular exercise of practical reasoning. The human subject plays the rather passive role of a "customer" who nevertheless enjoys a wide range of rights, and whose preferences and values it is the role of the policy-making institution to satisfy and accommodate. The human subject, so described, is apt for, and arguably in large measure the creation of, market-type institutions, and modes of decision-making that employ quantitative techniques of aggregation, weighing, and measuring. The degree of "fit" is to a large extent self-fulfilling. However, we have here appealed to philosophical reflection and to a variety of recalcitrant phenomena in order to suggest that this approach to decision-making, and the picture of the human subject that underpins it, are far from satisfactory. For what goes largely unnoticed is that markets, far from being institutions for revealing "raw" preferences, are in fact highly specialized institutions, *part of whose function is precisely to release us from a variety of well-grounded inhibitions that are normally in place.*

Our discussion has offered clues as to how decision-forming institutions might need to be modified to accommodate the alternative conception of the human subject that we have sketched. Appropriate alternative institutions will recognize that people often harbor conflicting and indeterminate values, and will behave in ways that are as much creative and experimental as predetermined, being responsive to reasons and to context. This suggests that institutions should be geared to inspire as much as to satisfy; they should also provide space for intersubjective and interactive scrutiny and challenge, and contexts for the building of trust and sharing of responsibility. They will seek to deserve, rather than manipulate, people's loyalty and commitment, and engage with their quest for meaning and continuity equally if not more than with their quest for satisfaction and security. They will acknowledge that "information" is rarely neutral – that it will invariably preempt certain approaches and suppress certain questions. What matters, then, is not what and how much "information" is provided, but how it is construed. They must be willing to invite reconstrual and rebuttal.

To the extent that values and preferences are incommensurable, aggregation is ruled out. To the extent that values and preferences stand in relationships to one another of amplification, endorsement, qualification, and cancellation, then any process of weighing conceived in quantitative terms is ruled out. Because the reason-giving force of values and preferences is contextual, it is virtually meaningless to construct justifications from the aggregation of preferences, and use these as the basis for determining public choice. Whether more sophisticated market instruments and adaptations of BCA techniques can overcome these difficulties of principle is a moot point. Experiments with what are termed "deliberative processes" – citizen juries, consensus conferences, and the like – seem at least to be pushing the institutional agenda in the right direction. But these too are beginning now to attract critical attention – and rightly so. Some of the criticism may be misguided inasmuch as it assumes the belief/desire model of the human agent that we have here rejected. If deliberative procedures are conceived simply as alternative ways of giving voice to and aggregating raw individual preferences, they may well be thought inefficient at best, and at worst open to all kinds of "contamination," in the form of manipulation, blackmail, and duress. But if that model is rejected, the criticism falls. Sounder criticism will focus on the effects of the uneven distributions of power, confidence, understanding, and oral skills that will apply to any given context of deliberation. Results may also be inconclusive, contradictory, or even irrelevant. What this points to is the maintenance of an overarching role for qualitative discrimination, by whomever exercised: democratic principles imply a right to be considered; but they do not imply a right to count, or to determine the outcome.

Conclusion

It matters what kinds of process and what kinds of institution are adopted for the articulation and expression of values in public policy, because:

1 Values are by their nature sensitive both to reasons and to context.
2 Values define us, give expression to autonomy and commitment, and are keyed into notions of identity and self-worth.
3 Values are as much created as revealed by the process or institution through which they receive expression.

The penalty for not developing institutions in which ethical and other deeply felt concerns can be properly voiced will be residues of grievance, mistrust, injustice, and guilt, which are as corrosive of the civic body as are pollutants in the natural environment.

References

Anderson, E. 1993: *Value in Ethics and Economics*. Cambridge, MA: Harvard University Press.

Aristotle 1999: *Nicomachean Ethics*, tr. Terence Irwin. Indianapolis: Hackett.

Burnyeat, M. 1980: Aristotle on learning to be good. In A. O. Rorty (ed.), *Essays on Aristotle's Ethics*. Berkeley: University of California Press, 69–92.

Chang, ?. (ed.). 1997: *Incommensurability, Incomparability, and Practical Reason*. Cambridge, MA: Harvard University Press.

Frey, B. S. 1994: How intrinsic motivation is crowded out and in. *Rationality and Society*, 6, 334–52.

Frey, B. S. 1997: A constitution for knaves crowds out civic virtues. *Economic Journal*, 107, 1043–53.

Gneezy, U. and Rustichini, A. 2000: A fine is a price. *Journal of Legal Studies*, 29, 1–18.

Holland, A. 1997: The foundations of environmental decision-making. *International Journal of Environment and Pollution*, 7, 483–96.

Hume, D. 1978 [1740]: *A Treatise of Human Nature*, ed. P. Nidditch. Oxford: Oxford University Press.

James, W. 1904: *The Will to Believe*. London: Longman.

Knight, F. H. 1922: Ethics and the economic interpretation. *Quarterly Journal of Economics*, 36, 454–81.

Lukes, S. 1997: Comparing the incomparable: trade-offs and sacrifices. In R. Chang (ed.), *Incommensurability, Incomparability, and Practical Reason*. Cambridge, MA: Harvard University Press, 184–95.

McNaughton, D. 1988: *Moral Vision: An Introduction to Ethics*. Oxford: Blackwell.

Plato 1993: *The Republic*, tr. R. Waterfield. Oxford: Oxford University Press.

Quinn, W. 1995: Putting rationality in its place. In R. Hursthouse, G. Lawrence, and W. Quinn (ed.), *Virtues and Reasons: Philippa Foot and Moral Theory*. Oxford: Clarendon Press, 181–208.

Raz, J. 1986: *The Morality of Freedom*. Oxford: Oxford University Press.

Raz, J. 1997: Incommensurability and agency. In R. Chang (ed.), *Incommensurability, Incomparability, and Practical Reason*. Cambridge, MA: Harvard University Press, 110–28.

Regan, D. 1997: Value, comparability and choice. In R. Chang (ed.), *Incommensurability, Incomparability, and Practical Reason*. Cambridge, MA: Harvard University Press, 129–50.

Scanlon, T. M. 1998: *What We Owe to Each Other*. Cambridge, MA: Harvard University Press.

Sober, E. 1995: Did evolution make us egoists? In *From a Biological Point of View*. Cambridge: Cambridge University Press.

Stern, P. and Dietz, T. 1994: The value basis of environmental concern. *Journal of Social Issues*, 50, 65–84.

Titmuss, R. M. 1970: *The Gift Relationship*. London: George Allen and Unwin.

Wiggins, D. 1980: Weakness of will, commensurability, and the objects of deliberation and desire. In A. O. Rorty (ed.), *Essays on Aristotle's Ethics*. Berkeley: University of California Press, 241–65.

3

The Ignorance Argument: What Must We Know to be Fair to the Future?

Bryan Norton

It is often noted that "sustainability" has come to "mean all things to all people," and indeed the term is used in many confusing ways; but we should not go too far in emphasizing the ambiguity of the term. It does, after all, have a clear, core meaning: sustainability is about the future and our concern toward it.

It thus seems reasonable to say that sustainability has to do with our inter-temporal moral relations and concerns our obligations to future generations. But do we have obligations to the future? It has been asked, for example, "Why should I care for posterity? What has posterity ever done for me?"[1] These questions acknowledge that our moral relations with the future are, inevitably, asymmetrical in an important sense. Obligations to the future cannot be of the standard, contractual variety; they cannot depend upon either individual prudence or assured reciprocality. And yet, despite the oddness of the obligations involved, most people show a significant concern for the future and for the impact of our choices on it. Opinion polls show that overwhelming majorities of people in modern democratic societies believe we should protect resources and natural wonders for the future. It is sometimes difficult to interpret this opinion and the sort of commitment it implies because people have different "mental models" of environmental change, and of the human role in it, but the survey data are very clear. Most people today, when asked to think about it, state that they prefer to live in a society that cares for the future and limits actions that are likely to have negative impacts on the future. Surely this widespread impulse is at least partially responsible for the widespread interest in, and acceptance of, sustainability as a public policy goal.

---

[1] O'Neill (1993) has an amusing explanation of the origin of these questions. My conclusions are similar to his, but the lines of our argumentation are quite different.

Intertemporal Ethics as Economics

Some societies are quite explicit and articulate about what they hope to bequeath to their offspring; but it is obvious that there will be a discrepancy between the actual bequest – the sum total of our impacts on subsequent generations, as they actually unfold through history – and the *intended* legacy that is spoken about, and sometimes planned for. I will refer to the "bequest package" that one generation leaves for the next – a bequest package is the sum total of accumulated capital, the technology, the institutions, and resources. The bequest package portends the sum total of *actual* impacts that one generation has on all subsequent generations. Given the prospective viewpoint of this chapter, discussion of proposed bequests will of necessity refer to intended bequests, not actual bequests.

However great the difference between the two, we must assume they are related: the intentions of earlier generations must at least matter in the formation of the actual bequest. Otherwise, we would not take intergenerational obligations seriously. Accepting an obligation to protect the future from negative impacts of our own decisions requires at least some faith in our ability to foresee those impacts and at least some of their consequences for future people, as well as some means to avoid identified unacceptable consequences. As we shall see shortly, the problem of ignorance about the future lies at the heart of one of the most interesting and lively theoretical debates about the meaning of sustainability.

Historically, people relied upon tribal elders and religious wisdom to guide actions with widespread and long-term consequences; today, science and scientific modeling are the favored means of projecting consequences into the future and guiding present-day decisions. Scientific models possess great potential for offering more precise (although not necessarily more accurate) predictions about actual impacts that will unfold in the future; they can also be used to generate "scenarios" of possible development paths a society might pursue. Yet our abilities to foresee the future and future impacts of our decisions remain limited. Every decision we make that affects the distant future is clouded in uncertainty and ignorance, and any successful account of our obligations to the future must deal with these problems. We simply cannot make a fair and reasonable assessment of the bequest package we prepare for the future unless we come to grips with the problem that, however sophisticated our scientific projections and models are, they will never have perfect foresight.

In this chapter, I explore intergenerational moral relationships in order to clarify the term "sustainability." I begin by addressing the economist's approach to the obligations to posterity, which will get us quickly to the issue of ignorance and the limits it may place on possible definitions and ideals of sustainability. Economists, much more than philosophers, have been seriously engaged in attempts to define sustainability and have succeeded in laying the foundations for the debate on intergenerational obligations. For example, many of those who explore obligations to the future take it for granted that discounting of future values will occur in evaluation processes; those who disagree usually propose alternative levels of discounting, different ways of computing the discount rates, or qualifications of the discounting practice. Economists have thus shaped the academic and policy discourse about intertemporal ethics. Their success in setting the agenda for the discussion of sustainability is due to an ingenious conceptual simplification – one that is based on an assumption regarding the role of knowledge in the process of choosing a fair bequest package. Accepting these assumptions, it is reasonable to consider our obligations to the future as a problem of finding a fair tradeoff between consumption today and foregone consumption today, in order to protect the consumption opportunities of future people.

I call this reduction of intergenerational equity to cross-temporal comparisons of welfare opportunities the "Grand Simplification" (GS). This simplification provides the theoretical foundation for the treatments of "sustainability" advocated by most mainstream economists and many philosophers. According to this view, sustainability is about affording each generation equal opportunity to enjoy undiminished welfare, by comparison to the welfare opportunities available to earlier generations. Philosophers who address the question of our obligations to the future have been supportive of such a simplification; they cede to the economists the intellectual territory where our actual obligations to future generations are to be determined. With the exception of the philosopher O'Neill (1993) and the economist Bromley (1998), there has been little resistance to the economizing of intergenerational morality. Here we will use the economists' version of this simplification to set the stage for a broader evaluation of the GS and its effects on both economics and philosophy.

Economists – especially growth theorists – have outlined a position that has come to be called *weak sustainability*. This view has been challenged by "ecological economists," who defend *strong sustainability*. Weak sustainability is based on the intuition that what we owe the future is to avoid actions that will make them *poorer* than we are in terms of opportunities to achieve welfare equal to ours. This is a weak requirement. Quite simply, each generation is required to maintain and pass on to their successors the economic capital they have inherited. No environmental goals should be given priority over other investments that have equal or greater expectation of return in terms of capital accumulation. Fairness is a matter of choosing the proper mix of investment and consumption.

Strong sustainability presents more stringent sustainability requirements. Strong sustainability theorists have defined a category of social value, called "natural" capital, which is not created by humans, and is deemed essential to the well-being of the people of the future. Strong sustainability requires the protection of natural capital (in addition to the more general requirement that each generation maintain the stocks of man-made capital) as a significant aspect of one generation's bequest to the next.[2] Strong sustainability requires an adequate and appropriate definition of natural capital, and I am not convinced such a definition can be provided (Holland, 1999; Norton, 1999). Strong sustainability theorists and others who wish to articulate more stringent sustainability requirements must provide good reasons to go beyond weak sustainability which, as we shall see, offers an initially attractive simplification of an otherwise perplexing moral problem.

[2] See, for example, Daly and Cobb (1989).

In a series of lectures and papers, Robert Solow has forwarded a view according to which "sustainability" can be fully defined, characterized, and measured within neoclassical economic theory (Solow, 1974, 1986, 1992, 1993). He argues that all we could possibly owe the future is that they be as well off, economically, as we are (weak sustainability). Solow's basic idea is that the obligation to sustainability "is an obligation to conduct ourselves so that we leave to the future the option or the capacity to be as well off as we are." He doubts that "one can be more precise than that." A central implication of Solow's view is that, while to talk about sustainability is "not empty, . . . there is no specific object that the goal of sustainability, the obligation of sustainability, requires us to leave untouched" (1993: 181). Sustainability, he says, is "a problem about saving and investment. It becomes a choice between current consumption and providing for the future" (1993: 183). Within this set of definitions, the future cannot fault us as long as we leave the next generation as able to fulfill their needs and desires as we have been. This position challenges strong sustainability and calls into question most of the environmentalists' programs, which include many more specific items of obligation. Solow's argument is worth examining in detail.

First, Solow dismisses a "straw man" of "absurdly strong sustainability" – the theory that we should leave the world completely unchanged for the future. "But you can't be obligated to do something that is not feasible" (1993: 180), he argues. Solow concludes that, since we cannot leave nature exactly as it is, there are no limits at all to the substitution of human-made wealth for natural resources. We should simply try to maintain an expanding total stock of capital. From this viewpoint, sustainability is a matter of balancing consumption with adequate investment, so that in the future there will be enough wealth to invest and to support people's desires to consume. On this view, financial assets, technology, labor, and natural resources are interchangeable elements of capital. Having dismissed the straw man, Solow asserts our total ignorance regarding the preferences of future people: "we realize that the tastes, the preferences, of future generations are something that we don't know about" (1993: 181). So, he argues, the best that we can do is to maintain a nondiminishing stock of capital. Solow says that "Resources are, to use a favorite word of economists, fungible in a certain sense. They can take the place of each other" (1993: 181). Because we do not know what people in the future will want, and because resources are intersubstitutable anyway, all we can be expected to do is to avoid impoverishing the future by overconsuming and under-saving. Provided that we maintain capital stocks across time, each generation will have an undiminished opportunity to achieve as high a standard of living as their predecessors. The ability of economies to find replacements for any scarce resource, if coupled with adequate economic capital for investment, will allow people of the future to fulfill whatever needs and wants they actually happen to have.

This argument reduces intergenerational obligations to fair tradeoffs across generations. The argument is appealing because it does seem wrong to expect earlier, poorer generations to sacrifice for the benefit of successors who turn out to be richer than they. According to this line of reasoning, provided that we manage tradeoffs between the present and the future such that people in the future have the opportunity to be as well off as we are, we will have achieved a fair tradeoff between obligations to the present and concerns for the future. Nothing more could be expected of us.

The GS is theoretically appealing because it cuts through a lot of confusing issues, and provides a clear and simple view of how to fulfill obligations to future generations. This weak sustainability approach is also practically attractive: it offers a criterion to judge whether or not a given society is "sustainable" – one can use economic growth and savings rates as a starting point for calculations – and therefore points the way for further research and experimentation. Even better, this sustainability requirement will no doubt be politically welcome – it suggests that changing over to sustainable development as a social goal involves little change from currently accepted goals of having an efficient, constantly growing economy, with a savings rate of greater than zero.

Environmentalists will – and should – be skeptical here; it seems as if Solow has swept away all concerns about environmental protection, submerging them under the more general calculations about savings and consumption. For him, the problem is to find good investments; investments that will grow rapidly enough to increase the wealth of the society even as they support increasing levels of consumption. However, Solow explicitly states that he does not suggest that other environmental policy goals are unimportant. He says:

> What about nature? What about wilderness or unspoiled nature? I think that we ought, in our policy choices, to embody our desire for unspoiled nature as a component of well-being. But we have to recognize that different amenities really are, to some extent, substitutable for one another, and we should be as inclusive as possible in our calculations. It is perfectly okay, it is perfectly logical and rational, to argue for the preservation of a particular species or the preservation of a particular landscape. But that has to be done on its own, for its own sake, because this landscape is intrinsically what we want or this species is intrinsically important to preserve, not under the heading of sustainability. Sustainability doesn't require that any *particular* species of owl or any *particular* tract of forest be preserved. (1993: 182)

This passage may seem to invoke the idea of "intrinsic value," as advocated by many environmental ethicists, who define intrinsic value as the value a thing has independent of humans, or at least independent of their motives and values. But this is surely an apparent similarity only: environmental ethicists who advocate intrinsic value in nature emphasize the *independence* of this value from human preferences, while Solow emphasizes their *dependence* on individual preference. For Solow, choices to save particular things should be based on the aggregated preferences of individuals to save something. I am not sure what to make of this rather bizarre terminological disconnect. Solow appears to avoid prescribing preferences – which is a priority for most nonanthropocentrists – and he seems to be denying that *moral obligations* are involved in these special commitments to places and species. I take it that this lack of moral status explains Solow's denial that decisions to protect special places or species are an element of sustainability. The decisions of earlier people to set aside places that they love would be in the nature of a gift in this line of reasoning, a free decision based on their own preferences, and not in any sense a fulfillment of a moral obligation, and thus no part of the obligation to live sustainably. Sustainability is a matter to be decided by the equations of growth-and-savings economists.

It is not surprising that Solow and other economists find the Grand Simplification attractive; it defines sustainability in a way that can be measured by the very techniques they can offer. It is more surprising that the few philosophers who have discussed obligations to the future have almost universally endorsed the GS. John Rawls (1971) argued that what we owe the future is to maintain a fair savings rate. Brian Barry has addressed obligations to the future several times, and in each case he advocates a position almost identical to Solow's (see, for example, Barry, 1989: 515–23). Other philosophical advocates of the GS include John Passmore (1974), the author of the first book-length treatment of environmental ethics, and the legal theorist Martin Golding (1981). So the GS has directly or indirectly shaped virtually the entire academic and public discourse about what sustainability is and what it means in normative and ethical terms.

Grandly Oversimplified?

I have raised several problems with the GS elsewhere (Norton, 1999), but here I want to concentrate on its foundational argument, ignorance, to show that the GS rests on shaky foundations. Solow's premise about the extent of our ignorance is stated so strongly as to defy plausibility. Solow seems to suggest that we have no idea what the future will be like, what challenges will be faced, and what people will want or need. Ignorance of what people will want wipes away any specific obligations beyond the next generation, because in a system of dynamic technological change we cannot identify any resources that will be crucial to tomorrow's production possibilities. It makes no sense to distinguish between unacceptable and acceptable use of resources – there are only "efficient" and "inefficient" ways to generate human welfare. As long as the future is wealthier, future people will have no right to complain that they were treated unfairly.

Solow's approach is about as simple as one can imagine, given that we address a problem as knotty as that of intergenerational equity. But simplicity, by itself, is not the goal of analysis. The theory should only be as simple as it can be, *given that it is adequate to the task at hand*. I have argued elsewhere that the problem of intergenerational equity is philosophically and morally complex (Norton, 1999). For example, there is a problem about the horizon of moral concern: How far into the future do our moral obligations extend? This can be called the *distance problem*.[3] There is also what I call the *typology of effects problem*: How can we

[3] Some economists – for example, Howarth and Norgaard (1995) and Bromley (1998) – have employed overlapping generational models and a sort of intergenerational "ratcheting" of concerns for the future, with each generation choosing the bequest for only the next generation. However, I find this solution raises as many questions as it answers. For example, how does such a solution support a fair policy regarding long-term radioactive waste?

decide which of our actions (and associated impacts) are governed by moral principles? And which effects are simply matters to be decided by market forces? If I cut down a mature tree and plant a seedling of the same species, it seems unlikely that I have significantly harmed people of the future, although there will be a period of recovery. As long as there are many trees left for others, my seedling will eventually replace my consumption, and no harm is done. If, on the other hand, I clear-cut an entire watershed and set in motion severe and irreversible erosion and siltation of a stream, I may have significantly harmed the future. I have irreversibly limited resources available to future persons, and I have restricted their options for pursuing their own well-being. It may be difficult to provide a theoretically justifiable definition, or a practical criterion, for separating cases of these two types.[4] But the ignorance argument sweeps this distinction away altogether and calls into question the goal of separating culpable from nonculpable actions that affect the future. If one is at all gripped by these philosophical problems, one should be left with a nagging concern. How can a question of fairness to future generations be decided without any attention to important moral issues such as the *distance* problem and the *typology of effects* problem?

Let us now look in more detail at Solow's use of the ignorance argument to reach the Grand Simplification. The Grand Simplification: (1) foreshortens our obligations to only the next generation, cutting off our obligations to the future entirely arbitrarily;[5] (2) assumes the fungibility of resources across uses and across time, denying the possibility of shortages or unfulfilled demands for natural resources; and (3) rules out *ex cathedra* the possibility that some courses of action will be economically efficient, and remain that way, but still impose uncompensated and noncompensable harms on future people.

The Grand Simplification is so grand because it resolves the distance problem by erasing all specific concerns for distant generations and because it sidesteps the typology of effects problem by assuming the fungibility of resources. Thus, all we need do is to avoid impoverishing the future by overspending and under-saving, which can be achieved simply by maintaining a fair savings rate. This simplification is based on no better foundation than an implausibly strong statement of the ignorance problem,[6] coupled with the implicit fungibility assumption according to which resources "can take the place of each other." These assumptions should not be considered as an empirical theory but, rather, as a proposed conceptual model for judging the sustainability of proposed policies and activities (Daly and Cobb, 1989; Norton, 1995). Solow's approach must therefore be examined not just for the verifiability of its assertions, but also with

---

[4] See Page (1983) for an excellent discussion of this problem, and a demonstration that criteria of economic efficiency cannot solve it.

[5] Presentism, so stated, might be thought to justify discounting. But many authors have argued that applying discounting across generations begs important moral questions. See Ramsey (1928), Pigou (1932), Parfit (1983), and Page (1988).

[6] See Callahan (1981), Kavka (1981), Page (1983), and Barry (1989) for convincing reasons that this strong ignorance premise is implausible.

respect to the appropriateness of its assumptions and conceptual commitments to the task of understanding what we owe the future. Operationally, these assumptions ensure the substitutability of resources, committing their advocate to the idea that, for any resource that may become scarce, there is a substitute that will stand in as an acceptable replacement for the loss.

This discussion has been mainly abstract and theoretical. It is appropriate to ask: What harm would result if we apply the GS, with discounting, to guide our resource use? This question cannot be answered conclusively unless we have some independent criterion for determining what we *should* do for the future. If we had such, the problem we are struggling with would already be solved. To avoid begging the question at issue, we must take a less direct approach to evaluating the consequences of the GS. For example, we can compare recommendations of the GS with intuitions of morally reflective individuals. Or, we can compare the bequest package that would be prepared by advocates of the GS with the package recommended by forward-looking environmentalists. In the remainder of this section, I will examine how well the GS does in comparison to our everyday intuitions and how much its recommendations diverge from the views of most environmentalists.

I have explored many examples of intuitions that conflict with the recommendations of the GS in Norton (1999). Here we will look at a few of them on our way toward a deeper examination of the role of ignorance in our choice of policies for long-term sustainability. If we take Solow's ignorance argument seriously, it seems as if we don't even know that people of the future will prefer clean water to water polluted with toxins. Perhaps future people will like to bathe in toxic wastes. This is not plausible; and one doubts that Solow meant his claim to be construed so generally. Surely there are some things we know about future preferences. What we know about them, together with extrapolations of risks from present behaviors, implies that some of our actions – such as careless storage of toxic wastes – could gravely harm people in the future. But if Solow admits that interventions in the market are necessary to rule out some specific actions as too risky, he must offer a criterion to identify these actions. The following question is unavoidable: Given what we reasonably believe about future tastes, what effects of our activities can be predicted to be benign and which are likely to be harmful? But once we raise this question, the question about the typology of effects immediately returns to center stage. And then the Grand Simplification unravels. We are back to trying to figure out what we owe, to how many generations, with a knowledge base that contains some near certainties and a great deal of uncertainty.

The generic recommendations of advocates of the GS diverge in important ways, also, from the commitments of environmentalists, who believe we owe specific things to the future, such as clean water, biodiversity, and natural areas. Wilderness protection and protection of special places such as Chesapeake Bay highlight this disagreement between Solow's theory and the intuitions of environmentalists, a disagreement that extends far beyond the measures of consumption and welfare opportunities. Here, we must return to Solow's discussion of why we save particular species and landscapes. As noted in the first section,

Solow says that we may save particular species and places because we love them for their own sakes. I assume that he means by this that we should save those species and places that best fulfill the aggregated preferences of all of today's consumers (not that we must save all species and places that someone does in fact love). Notice that Solow shifts attention from what people in the future will want toward what people in the present care about. I agree with this strategy; but I would not follow Solow in reducing all obligations beyond weak sustainability to aggregated preferences for saving certain "preferred" landscapes and species. To see the difference it is useful to compare Solow's attitude about future preferences to that of the environmentalists. Solow claims that not only do we not know what the future will want but that "to be honest, it is none of our business" (Solow, 1993: 182). A comparison of this passage with attitudes of environmentalists tells us a lot about the difficult relations between mainstream economists and environmentalists.

When environmentalists assert an obligation of the present generation to protect special places from severe degradation, they may well make assumptions about what will be valued in the future. They might, for example, assume that people in the future will greatly value these special places. But environmentalists also believe that people in the future *should* value these special places. Imagine that our generation, through conscientious effort and some sacrifice, succeeds in protecting many special places. Further, suppose that our children's generation continues the protection, but that the next generation prefers development over preservation and systematically destroys the natural legacy we have left them. If we could somehow learn that our grandchildren or great-grandchildren will desecrate the heritage we so carefully preserved for them, I submit that we would not, as passively as Solow says, accept this as none of our business. On the contrary, we might well increase efforts to educate today's population and to build lasting institutions that will perpetuate our deeply held values and ideals. Environmentalists accept responsibility to protect places – and in doing so they also accept responsibility to foster a *commitment to protection*. There is a paternalistic streak in environmentalism. Environmentalists hope to save the wonders of nature, but they also accept a responsibility to perpetuate a love and respect for the nature they have loved enough to protect (Sagoff, 1974; Sax, 1980).

I noted above that environmentalists would balk at Solow's ignorance argument. It is now clear that their disagreement with Solow is not based mainly on a belief that they are able to *predict* what people in the future will in fact prefer. It is, rather, that they *accept moral responsibility* for inculcating certain values, and for ensuring that those values are perpetuated in future generations. This analysis suggests that wilderness areas and other natural wonders are not valued by present and future preservationists simply as opportunities for preference-satisfaction and for welfare gain.

Two directions are open to us to explore the preservationist values. One possibility is to attribute intrinsic value directly to nature or its elements. This approach has led mostly to futile discussions of what has and what lacks intrinsic value, and will not be discussed here at any greater length. The other alternative is to understand the value placed on special places without either positing intrinsic value in nature or falling back upon the requirements of weak sustainability only. One can argue that there are ways in which we could harm future people *even if they are as well off as we are*. We can define a class of harms to the future people that would leave them worse off than they would have been had these impacts been avoided. If such a class can then be defined independently of effects on productivity, on capital accumulation, and on social wealth, we would have identified a category of "noneconomic" obligations to the future. These are the values that people are referring to when they say that certain possessions or experiences are "priceless." Loss of such a value can represent a harm, despite any level of compensation represented by wealth or income.

While it would take more space than is available here to make a conclusive case that such obligations exist, the fact that such a class can be coherently defined shows that there is a possibility of developing a stronger notion of sustainability, and of doing so in at least partially noneconomic terms. Having already admitted that weak sustainability may state a necessary condition of sustainable living, I now explore whether there may be other obligations that would govern our behavior toward the future. It seems reasonable to seek such long-term obligations not in concerns to fulfill future consumer preferences, but in our debt to prior generations who have developed our society, created our culture, and established a community with reasonably democratic and noncoercive institutions, and formed a distinctive relationship to a geographical place.

Sustainability and Community-Based Obligations

Many environmentalists believe it is possible to harm the future in a way that could not be morally repaired, even if those in the future turn out to be wealthier than we are. I suggest we explore the idea of noncompensable harm by reference to violations of obligations to a multigenerational community. These include obligations not to destroy the natural and cultural history of a "place" in which humans and nature have interacted to create an organic process, a process that can be understood multigenerationally. The communitarian, unlike the welfare economist, sees goods beyond individual ones, rejecting the economist's model of decision making as based solely on aggregation of the individualistic values of *Homo economicus*. The communitarian recognizes also community-based values that connect a human community to its natural context.

Environmentalism can be understood within the broadly communitarian political ontology of the conservative philosopher Edmund Burke, who defined a society as "a partnership not only between those who are living, but between those who are living, those who are dead and those to be born" (1910: 93–4). What environmentalists need to add to Burke's political community is a stronger sense of human territoriality and a more explicit recognition that both our past and our future are entwined with the broader community of living things and eco-physical systems that form the habitat and the context of multigenerational human communities. Nature protectionists do not, as citizens in Burkean communities, evaluate special places as possibilities for present or future consumption, but rather as shrines, as occasions for present and future people to recollect, and stay in touch with, their authentic natural and cultural history. This is a history of humans as evolved animals, and also as cultural beings who have evolved culturally within a particular natural setting. These are cultural beings who cannot deny their wild origins (Leopold, 1949: 199–200; Thoreau, 1960: 144).

Here, I think, we have reached the nub of the matter. What holds the supporters of the GS together is methodological individualism, the view that *the* good must be an aggregation of *individual* goods. Countering this, I recommend that noneconomic obligations to the future be considered as community-based goods (O'Neill, 1993). We have these obligations because, as members of a community and a culture, we benefit from sacrifices and investments made by members of prior generations. While these benefits include economic goods, they are not reducible to such because they include the political and cultural practices that give meaning and continuity to the culture. Adam Smith (1776) talked of these practices and sensibilities as the "bonds of sympathy" and considered them an essential foundation of economic life. These bonds cannot be valued in economic terms. To do so is to commit an egregious category mistake (Sagoff, 1988: 92–4). These concerns are best understood by paying attention to the moral and cultural sentiments of persons and communities, and by emphasizing the ways in which these sentiments form an essential part of a person's personal and community identity.[7]

Some obligations to the future are obligations to build a community and culture that lasts. We would have failed the future if we fail to develop institutions, ideas, and practices that create lasting communities in real places, communities with the moral and institutional strength to protect special places as symbols of the natural history and emergent culture of the community. Failure to do so, I believe, could leave people of the future worse off than we are in an important respect, even if they are vastly wealthier than we are. To see how this might be true, suppose our generation systematically converts all old-growth forests and wilderness areas to farming and mining, producing one form of "wealth," but making it impossible for future persons to experience unspoiled wilderness or other natural places. If they are wealthy, they can perhaps provide themselves Disney-type facsimiles of wilderness. As long as they have adequate income to afford such substitutes, they will have been adequately compensated – according to advocates of weak sustainability – for the loss of natural resources. Environmentalists, of course, would reject this reduction of all value to opportunities to

---

[7] See Holland and O'Neill (1996) and Holland and Rawls (1993) for a useful discussion of the inseparability of cultural and ecological ideals. Also see Ariansen (1997). Ariansen suggests that some choices we make with respect to protecting our environment represent "constitutive" values. Loving and protecting special places and special features of a place, on this view, may be constitutive of a person's sense of self and of community membership. Careless destruction of these special features, correlatively, might be considered a kind of "cultural suicide." Ariansen's insight provides one interesting direction for the explication of what I call noncompensable harms.

produce and consume in the future. We can harm the future by failing to create and maintain a culture and a community respectful of its past, including both the human and natural history of the common heritage.

Successful protection of wilderness and special places such as the Chesapeake Bay requires a successful transmission of love, respect, and caring for these places to the persons who will live in subsequent generations – including a sense of moral obligation to continue to protect them. Solow, who is committed to the individualistic, utilitarian view of mainstream welfare economics, sees the value of an object as identical to its ability to fulfill preferences that people happen to have as individual consumers. Rejection of this individualistic value theory is at the very heart of the environmentalists' position. They not only work to protect natural resources and places in the present, but they also attempt to project their values into the future. The questions of how people come to have, and to change, their preferences, and of whether we can judge some preferences morally superior to others, are interesting and important questions for understanding intergenerational equity issues. But these questions have not been given attention by economists, because such questions are exogenous to the discourse of welfare economics.[8] Accordingly, the vocabulary of the economist models all values as fulfillments of individual preferences and cannot express a central aspect of environmentalists' concern for the future.

As noted above, not all economists agree with Solow's simplistic concern with capital and wealth accumulation, so there is a basis for developing an alternative understanding of sustainability from within economics. For example, Bromley (1998) has argued, congruent with the argument developed here, that identifying a rational bequest is more a matter of *what* is left as a legacy for the future than it is a question of *how much* is left. Bromley explicitly favors looking to the present – to commitments of present people – as the source of guidance in choosing sustainability goals and principles, and not to speculations about the opportunities of the future to consume. Bromley's approach, like the one adopted here, must rest the matter of intergenerational morality on political will and on institutional development, rather than on counting wealth and "fungible" capital. The development of a worthy bequest for subsequent generations will require us to address hard moral questions about ourselves and the nature of society we want to build and transmit.

From within Solow's perspective, making moral judgments about the preferences of future people is irrelevant to sustainability. The advocate of weak sustainability cares only about the economic means available to satisfy whatever preferences future people happen to have. But for the environmentalist, it makes sense to say that those people in the future, who have lost all interest in nature, are worse off in ways that have little to do with their ability to fulfill their actual preferences. This claim may be controversial. Solow would no doubt argue that it is meaningless and that it is an advantage of his value calculus that claims such

[8] I have argued elsewhere that the economists' theoretical/methodological commitment to "consumer sovereignty" is a key source of disagreement between environmentalists and economists. See Norton (1994) and Norton, Costanza, and Bishop (1998).

as this fall by the wayside. But, environmentalists do exhibit commitment regarding the values that future people express and act upon, and this commitment cannot be expressed in the utilitarian calculus of economists. Can we help the environmentalists to make sense of their claim as a *moral* claim? If we can do so, we will have gone a long way toward clarifying the meaning of "noncompensable harm," and toward a stronger sense of sustainability.

What We Owe the Future

The formulation of intergenerational moral problems as utility tradeoffs, and as responding – in the manner of Solow, Passmore, and others – to the preferences of future persons, dooms any hopes of specifying stronger sustainability principles. If specifying our obligations to the future depends upon predicting in detail what individuals in the future will want or need, then assertions of obligations to the future will, at best, be plagued by unavoidable uncertainties. If we can be fair to the future only if we can predict their needs and preferences in detail, then there will always be an impossible task at the heart of all specific ("strong") sustainability requirements. From this perspective, the reduction of sustainability to weak sustainability – the reduction of future obligations to determining a fair savings rate – is simply a figment of the assumptions introduced in order to characterize the moral problem in utilitarian and economistic terms. Prior theoretical commitments to utilitarianism and economic operationalization of intertemporal welfare comparisons determine the contours of the playing field on which intergenerational obligations are discussed and determined. By insisting that intergenerational moral obligations be measured in terms of comparisons of aggregated welfare, utilitarians define intergenerational fairness so as to require information that cannot be available at the time when crucial decisions about what to protect must be made. The collapse of sustainability into weak sustainability on the basis of ignorance is preordained by the chosen theoretical scaffolding.

This outcome can be traced to the utilitarian dogma that normative questions must be construed as empirical questions with empirical answers. This dogma brings the question of *prediction* of wants, needs, and demands center stage in discussions of our obligations to the future; and it is this dogma that undermines any attempt to specify stronger sustainability requirements within the broadly utilitarian framework of analysis. At its deepest level, the Grand Simplification rests not upon the *fact* of our ignorance about future values but, rather, on a deep and unquestioned commitment to reduce all moral questions to descriptive questions – to questions that can be fully resolved on an empirical basis. The commitment of economists to the empirical resolubility or dismissal of moral questions pushes them toward a commitment to measuring and comparing quantities of welfare across time. This tendency puts extraordinary weight on our ability to *predict* future values and preferences. Furthermore, this commitment renders the analytic framework of welfare economics unable to express the core ideas of environmentalists.

There is a name for the mistake committed by those who seek to construe sustainability as comparisons of measurable quantities of well-being over time: it has been called the "descriptivist fallacy" by J. L. Austin (1962: 5–7). Austin argues that many of our sentences that look like ordinary statements have purposes other than to describe. As examples, Austin mentions "I do" when uttered in the context of a marriage ceremony; "I name this ship the *Queen Elizabeth*," while striking the ship's bow with a bottle of champagne; and "I bequeath this watch to my brother," in the context of a will. He argues that to utter these sentences is not to *describe* the doing of what is being done but, rather, to *do* it. Austin proposes that we characterize such uses of language as "performatives," and that there can be many types of performatives, including "contractual" and "declaratory" ones. Later, Austin says: "A great many of the acts which fall within the province of Ethics . . . have the general character, in whole or in part, of conventional or ritual acts."

Reflecting on the views of environmentalists, we found that they embrace a *commitment*, not only to save special places, but also to create and sustain institutions and traditions necessary to carry on the commitment indefinitely. These acts must include the creation of a place-based literature and narratives, as well as public and private "trusts" set up to secure, for example, habitat for indigenous species. All of these actions signal commitments to continuity between the past and the future; they are best understood in Austin's sense as "performatives." They are founded on the commitments that a community makes to continuity with its past, to its natural and cultural histories, and to a future in which its roots in nature are revered, protected, learned from, and cared for. These commitments are, one might say, "community performatives," and represent a community-based commitment to love and protect one's natural as well as one's social history.

One might ask whether my argument has not simply circled back to Solow's implication that the urge to protect loved species and special places simply reflects aggregate preferences of the society. In what way is my final position on sustainability "stronger" than Solow's? In my approach to the problem, each generation has an obligation to contribute to the cultural fabric of the community of which it is a member. And, if those communities and their members feel that their natural as well as their cultural history embodies and expresses their deepest values – as passed down from earlier generations and augmented in the present – then they have an obligation to create institutions and cultural supports for those values. To do less would be to sever important cross-generational ties, to leave the future with a diminished connection with its past, and a poorer cultural legacy to pass on to their children. This, I have argued, could impose upon the future a noncompensable harm – they would be worse off for our carelessness, regardless of their opportunity to fulfill, in the absence of cultural and moral guidance from us, whatever preferences they happen to have. Unlike Solow's calculations, the search for a stronger sense of sustainability must take place in the present, and it must search the hearts and minds of humans who are committed to building a community that reflects both cultural and natural ideals. Ignorance of people's future preferences need not deter this search. Further, my approach demands a stronger sense of community; a sense of community that recognizes bonds to an ongoing project, as well as obligations in the distribution of opportunities to consume.

Applying Austin's idea, and based on the analysis of this chapter, a new way of thinking about intergenerational morality emerges. If we see the problem as one of a community making choices and articulating moral principles – a question of which moral values the community is willing to commit itself to – the problems of ignorance about the future become less obtrusive. On the more communitarian approach suggested here, lack of knowledge of the detailed tastes of future people provides no real threat to the intellectual and practical task of specifying a fair bequest package for the future, because deciding what is owed the future is not only about fulfilling the needs that people in the future will in fact have, but it also has to do with affecting what those preferences will be through the creation of communally validated social values. The question at issue is a question about the present; it is a question of whether the community will, or will not, take responsibility for the long-term impacts of its actions, and whether the community has the collective moral will to create or contribute to a community that represents its own distinct expression of the nature–culture dialectic as it emerges in a place. Will members of the community consciously and conscientiously choose and implement a bequest package – a trust or legacy – that they will pass on to future generations? We do not then ask what the future will want or need – we ask by what process a community might specify its legacy for the future.

If one wishes to study such questions empirically, there is information available. One might, for example, study how communities engaged in landscape or ecosystem management achieve, or fail to achieve, consensus on environmental goals and policies. While empirical studies such as these may contribute to community-based environmental management, I suggest that the foundation of a stronger sustainability commitment lies more in the community's articulated moral commitments to the past and to the future than in any *description* of consequences for individual welfare. That we owe something beyond mere riches to the future follows from the fact that we inherited far more than mere riches from the past. But whatever information we need to answer the question of what, in particular, we owe the future must be discoverable in the present. It will have to do with what is important to our culture today – not with what people in the future will prefer.

This basic point makes all the difference in the way information is used in defining sustainability, and it changes the way we should think about environmental values and valuation. If the argument of this chapter is correct, then the problem of how to measure sustainability, while important, is logically subsequent to the prior question of commitment to preserving a natural and cultural legacy. So, we face the prior task – and I admit it is a difficult and complex one – of developing processes by which democratic communities can explore their common values, their differences, and choose which places and which ecosystems and which traditions will be saved, achieving as much consensus as possible, and continuing debate about differences. Commitments, made by earlier generations, represent the voluntary, morally motivated contribution of the earlier generation to the ongoing community. While choosing measurable indicators is logically subsequent to commitment to moral goals, the tasks of choosing measurable indicators can, and must, proceed simultaneously with the articulation of long-term environmental goals. It cannot be otherwise, because the choices that are made by real communities regarding which indicators are relevant to their moral commitments represent, in effect, an operationalization of moral commitments. The task of choosing community values similarly cannot be sharply separated from the specification of certain indicators that would track the extent to which actual choices and practices achieve those commitments. The specification of a legacy, or bequest, for the future must then ultimately be a political problem, to be determined in political arenas. The best way to achieve consensus in such arenas is to involve members of communities in an articulation of values, in a search for common management goals, and to include in that process a publicly accountable search for accurate indicators to correspond to proposed management goals.

The advantages of this shift in perspective are now evident: this approach suggests that the key terms, "sustainable" and "sustainable development," are not themselves abstract *descriptors* of states of societies or cultures, in general but, rather, refer to many sets of commitments of specific societies, communities, and cultures to perpetuate certain values, to project them into the future, and to build a strong sense of community and a respect for the "place" of that community, complete with institutions to support these values. The problem of how to measure success and failure in attempts at living sustainably is now the problem, for each community, of choosing a fair natural legacy for the future, democratically, and then operationalizing these commitments as concrete goals to be measured by democratically agreed-upon indicators. The problem of tradeoffs is still a key issue, but it is more manageable because it is no longer dominated by the constraints imposed by our ignorance about the future. The tradeoffs problem no longer appears as a problem of comparing aggregated welfare at different times, but as a problem of allocating resources to various, sometimes competing, social goals.

Here, it is undeniable – as the economists will be quick to point out – that, ultimately, people in the present must balance their concern and investments for the future against the needs of today. There are situations in which setting aside special places will compete with other values. But now the question is transformed. If we think less about intertemporal tradeoffs, and work toward creating fair and lasting institutions – institutions that sustain and provide fair access to resources now and in the future – these practices might be the most effective first steps that can be taken today. If we see the problem as one of commitment of today's people not to see certain of their values and commitments eroded, the fact of our (partial) ignorance of future desires and needs – while a limitation in some ways – is not really relevant to the environmentalists' case. They must make the case that, to the extent the community has committed itself to certain values and associated management goals, these goals are deserving of social resources and "investments" in the future. The task for the environmentalists is a daunting one, given the competing demands upon society's limited resources. To the extent that a community and its members see the creation of a legacy for the future as a contribution to an ongoing dialectic between their culture and its natural context, and to the extent that they accept responsibility for their legacy to the future, they have embraced a commitment that gives meaning and continuity to their lives. To create the institutions necessary to accomplish this goal, one must start today, and one must work with the information that is available today. The recommendation that emerges from this chapter is that the moral resources to begin, and carry forward, this task will be found in strong communities, not in attempts to guess and anticipate what people in the future will want.

References

Ariansen, P. 1997: The non-utility value of nature: an investigation into biodiversity and the value of natural wholes (translation from Norwegian text). In *Communications of the Norwegian Forest Research Institute (Meddelelser fra Skogforsk)*. Aas: Agricultural University of Norway, 47.

Austin, J. L. 1962: *How to Do Things with Words*. Oxford: Oxford University Press.

Barry, B. 1989: *Democracy, Power, and Justice*. Oxford: Clarendon Press.

Bromley, D. W. 1998: Searching for sustainability: the poverty of spontaneous order. *Ecological Economics*, 24, 231–40.

Burke, E. 1910: *Reflections on the Revolution in France*. London: Dent.

Callahan, D. 1981: What obligations do we have to future generations? In E. Partridge (ed.), *Responsibilities to Future Generations*. Buffalo, NY: Prometheus Books, 73–85.

Daly, H. and Cobb, J. 1989: *For the Common Good*. Boston: Beacon Press.

Golding, M. 1981: Obligations to future generations. In Ernest Partridge (ed.), *Responsibilities to Future Generations*. Buffalo, NY: Prometheus Books, 61–72.

Holland, A. 1999: Sustainability: should we start from here? In A. Dobson (ed.), *Fairness and Futurity*. Oxford: Oxford University Press, 46–68.

Holland, A. and O'Neill, J. 1996: The integrity of nature over time. *Thingmount Working Paper 96-08*. Lancaster, UK: Department of Philosophy, Lancaster University.

Holland, A. and Rawls, K. 1993: Values in conservation. *Ecos*, 14, 14–19.

Howarth, R. B. and Norgaard, R. B. 1995: Intergenerational choices under global environmental change. In D. W. Bromley (ed.), *Handbook of Environmental Economics*. Oxford: Blackwell, 111–38.

Kavka, G. 1981: The futurity problem. In E. Partridge (ed.), *Responsibilities to Future Generations*. Buffalo, NY: Prometheus Books, 113–15.

Leopold, A. 1949: *A Sand County Almanac*. Oxford: Oxford University Press.

Norton, B. G. 1994: Economists' preferences and the preferences of economists. *Environmental Values*, 3, 331–2.

Norton, B. G. 1995: Evaluating ecosystem states: two paradigms of environmental evaluation. *Ecological Economics*, 14, 113–27.

Norton, B. G. 1999: Ecology and opportunity: intergenerational equity and sustainable options. In A. Dobson (ed.), *Fairness and Futurity*. Oxford: Oxford University Press, 118–50.

Norton, B. G., Costanza, R., and Bishop, R. 1998: The evolution of preferences: why "sovereign" preferences may not lead to sustainable policies and what to do about it. *Ecological Economics*, 24, 193–212.

O'Neill, J. 1993: Future generations, present harms. *Philosophy*, 68, 35–51.

Page, T. 1983: Intergenerational justice as opportunity. In D. MacLean and P. Brown (eds.), *Energy and the Future*. Totowa, NJ: Rowman and Littlefield, 38–58.

Page, T. 1988: Intergenerational equity and the social rate of discount. In V. K. Smith and J. Krutilla (eds.), *Environmental Resources and Applied Welfare Economics*. Washington, DC: Resources for the Future.

Parfit, D. 1983: Energy policy and the further future: the social discount rate. In D. MacLean and P. Brown (eds.), *Energy and the Future*. Totowa, NJ: Rowman and Littlefield, 166–79.

Passmore, J. 1974: *Man's Responsibility for Nature*. New York: Charles Scribner's Sons.

Pigou, A. C. 1932: *The Economics of Welfare*. London: Macmillan.

Ramsey, F. 1928: A mathematical theory of saving. *Economic Journal*, 38, 543–59.

Rawls, J. A. 1971: *A Theory of Justice*. Cambridge, MA: Harvard University Press.

Sagoff, M. 1974: On preserving the natural environment. *Yale Law Journal*, 84, 205–67.

Sagoff, M. 1988: *The Economy of the Earth*. Cambridge: Cambridge University Press.

Sax, J. 1980: *Mountains Without Handrails*. Ann Arbor, MI: University of Michigan Press.

Smith, A. 1976 [1776]: *The Theory of Moral Sentiments*. New York: Garland.

Solow, R. M. 1974: The economics of resources or the resources of economics. *American Economic Review Papers and Proceedings*, 64, 1–14.

Solow, R. M. 1986: On the intergenerational allocation of natural resources. *Scandinavian Journal of Economics*, 88, 141–9.

Solow, R. M. 1992: *An Almost Practical Step Toward Sustainability*. Invited lecture on the occasion of the 40th anniversary of Resources for the Future, Washington, DC.

Solow, R. M. 1993: Sustainability: an economist's perspective. In R. and N. Dorfman (eds.), *Economics of the Environment: Selected Readings*. New York: W. W. Norton, 179–87.

Thoreau, H. D. 1960: *Walden*. New York: The New American Library.

4

Benefit–Cost Considerations Should be Decisive When There is Nothing More Important at Stake[1]

Alan Randall

Benefit–cost analysis (BCA) is often in the news, lately. We hear more and more proposals for BCA of regulatory actions, with a presumption that those that do not pass the benefit–cost test are, to say the least, suspect. Opposing this suggestion are those who claim, for a variety of reasons, that benefits and costs are not appropriate considerations when deciding on policy concerning human health and safety, environmental protection, and quality of life.

Calls for routine use of BCA are not new. BCA already enjoys a considerable role in public affairs. In the past half-century, benefit–cost analysis has evolved from a relatively crude financial feasibility analysis for capital-intensive public works (water resources and transportation projects were early applications in several countries) to a sophisticated and comprehensive application of the economic–theoretic principles of welfare change measurement to evaluate all manner of projects, programs, and policies. In the US, various federal Executive Orders have required BCA for a broad range of regulatory initiatives, while well-entrenched executive practice in many countries routinely considers benefits and costs for a considerable set of activities undertaken by national and local governments. On the other hand, the public role of BCA has been explicitly limited in many countries; for example, US legislation forbids a BC test for regulations protecting public health.

This substantial and controversial public role for BCA raises an obvious question: What justifies attention to benefits and costs in public affairs? One useful way to pose the question is: Does a benign and conscientious public decision-maker have a duty to consult an account of benefits and costs, as economists understand the terms benefits and costs (Copp, 1985)? However, Copp's "benign and conscientious public decision-maker" has a rather old-fashioned progressive ring to it, in these

---

[1] This chapter addresses a subject discussed in Randall (1999), revising the issues and extending the argument in several ways.

days when pluralism and participatory processes are all the rage. So it might be appropriate to offer an alternative statement of our fundamental question: Would (or should) a society of thoughtful moral agents agree to take seriously an account of benefits and costs?[2]

Economists' Justifications

Waste avoidance

When called upon to justify the systematic use of BCA in public decision processes, economists are likely to start talking about the need to impose a market-like efficiency on the activities of government (Arrow et al., 1996). After all, efficiency is simply the avoidance of waste, and who could be seriously in favor of waste! However, this justification is not as convincing to citizens at large as it is to economists.

First, a case has to be made that the efficiency of markets is in fact good for society. Perhaps we can do little better than Jules Coleman (1987), who has argued that the virtues of market institutions (including, but not limited to, their efficiency properties) makes them broadly acceptable for taking care of those kinds of human affairs that are not especially contentious (perhaps he meant the kinds of things that can be handled consensually by arms-length transactions), but that political institutions are required to deal with the really contentious issues of public concern. In other words, the justification for market processes applies to those human concerns that remain after the seriously contentious issues have been resolved, at least provisionally, by the institutions of government.

Having found some virtue in markets for handling some sets of human concerns, it is then necessary to argue that society ought to require market-like efficiency in the remaining undertakings that have been assigned for good reasons to government. Mark Sagoff (1981) is most vigorous in rejecting this argument. He asserts that it is a simple category mistake to inquire about the efficiency of a governmental undertaking: government is exactly that institution that human societies invoke when they choose, for their own good reasons, not to be efficient. It is easy to play this argument for cheap laughs ("Of course! What better institution than government, if the goal is to be inefficient!"), but Sagoff's point is not entirely frivolous. Efficiency is a harsh discipline, and one that in practice tends to reinforce the distributional status quo; and it is by no means clear that society ought to impose that discipline on everything that it does.

A filter for rent-seeking

Perhaps Sagoff is too naïve in his implicit assumption that government is invoked always for socially benign purposes. The public choice tradition, much more cynically, imagines individuals and interest groups attempting to use the power of government in service of their own purposes, restrained only by their private calculations of gains and losses. In such a public environment, BCA

---

[2] In framing the question in these ways, there is an implicit maintained assumption that it is feasible to produce reasonably accurate accounts of benefits and costs for a broad array of potential public actions. Another way of making the same point is to state directly that this essay simply is not addressed to difficulties in practical implementation of BCA.

provides a filter for negative-sum rent-seeking proposals – those that would cost the rest of society more than the beneficiaries would gain – and is often justified by economists for that reason.

This justification, however, is closely related to the waste avoidance justification. The idea of negative-sum rent-seeking merely recognizes that even waste benefits someone, and the beneficiary is motivated to do what can be done to encourage it. Not surprisingly, this justification encounters many of the same difficulties that face the waste avoidance justification. One person's public "good thing" is another's boondoggle, and the economist offers only the concept of efficiency to sort them out.

Welfarism

When asked to defend, in the standard vocabulary of moral philosophy, an efficiency test for government undertakings, economists are likely to develop an argument premised on the ethic of welfarism: the goodness of an individual life is exactly the level of satisfaction of the individual's preferences, and the goodness of a society is a matter only of the level of satisfaction of its members.[3] From these premises, economists have developed, invoking various assumptions and restrictions as necessary and convenient, the whole apparatus of welfare change measurement, of which BCA is the direct practical implementation. BCA is, then, an empirical test of whether proposed public actions would increase preference satisfaction.

The potential Pareto-improvement criterion

To implement the principle of welfarism, it is necessary to invoke some specific conventions about how things are to be valued. Among welfare economists, a consensus definition of BCA has at last emerged: BCA is an empirical test for potential Pareto-improvements (PPIs). The PPI test evaluates a proposed change by asking whether the amount that those who stand to gain would be willing to pay to get the change exceeds the amount of compensation that would induce those who stand to lose to consent to the change.[4] It implements welfarism by evaluating goodness to the individual in terms of buyer's best offer and seller's reservation price (which reflect endowments as well as preference satisfaction, a point examined at some length later in this essay), and goodness to society by adding up the resulting benefits and costs across individuals, without reference to distributional concerns.

Ethical Objections to Welfarism

Consider the status of welfarism in Western ethical theory. Welfarism is most readily understood as a kind of consequentialism: right action is whatever

---

[3] This definition follows Sen (1989). According to Kagan (1998), current usage among philosophers defines welfarism more narrowly – as evaluating welfare by the Benthamite utilitarian welfare function – and thus ignoring distributional concerns.

[4] The convention that BCA is a test for PPIs resolves an objection raised by Copp (1985): that BCA is not rigorously defined, so that we cannot know exactly what we are buying into. We can write down exactly what a tool must do in order for it to be a valid PPI test.

produces good consequences, and consequences are evaluated according to their contribution to welfare. However, the welfarist justification of an efficiency test is unlikely to be entirely satisfactory to consequentialists as a group – because many of them would insist that all manner of consequences not readily reduced to welfare are nevertheless worthy of consideration[5] – and even to many utilitarians, who consider preference satisfaction an important consideration but not strictly the only concern in evaluating the goodness of a society or an individual life. Furthermore, consequentialism is itself a particular version of axiology, the theory that goodness is a matter of value (Vallentyne, 1987). Not all axiologists would want to confine considerations of value to consequences alone.

Axiology is only one of the foundational ethics in the Western tradition. Philosophers frequently identify two foundational ethics, the second being deontological: goodness is whatever emerges from right action, so that the ethicist's task is to judge not value but the rightness of actions.[6] Economists often find it useful to distinguish two classes of deontological ethics: Kantianism, which defines right action as that which is obedient to moral duties derived ultimately from a set of universal moral principles; and contractarianism, in which right action respects the rights of individuals. These positions are both deontological, because the justification of Kantian moral imperatives and of individual rights requires appeal ultimately to some asserted principle.[7] Economists are comfortable with the contractarian perspective, which has become quite familiar and congenial, for example, in the justification of voluntary exchange. However, Kantian ethics – with its insistence that moral and prudential reasoning are quite distinct, and that universal moral imperatives can be found from which to deduce rules for action in practical situations – is quite bewildering to many economists, although not to many educated lay-persons.

Viewed in the light of the major foundational ethics, welfarism is found wanting. It is at best incomplete as an axiological moral theory, while deontological theories are premised on the belief that axiological (value) theories of the good are seriously wrong about some important moral questions. In the context of environmental projects and policies, one glaring weakness of welfarism is its inability to take seriously the concept of intrinsic value – that some things have value independent of any satisfaction they might provide a user or observer – an idea that is unexceptional to many nonwelfarists. Any justification

---

[5] For example, MacIntyre (1979) complains that BCA is committed to the commensurability of diverse values, an argument that includes but also extends beyond consequentialist values.

[6] Here, I take my cue from Rawls (1971), who – in order to contrast his approach with consequentialism – asserted flatly that justice is fairness, and fairness is whatever results from fair processes. Kagan (1998), who is concerned not with intellectual roots but with what philosophers are arguing about today, defines as deontological any ethical position that would impose constraints on the pursuit of axiological good.

[7] Thus, while Kant insisted that reason was sufficient to establish categorical imperatives (universal moral duties), he was unable to defend convincingly his claim that "not lying" should be one of them. Similarly, the attempt to ground contractarian ethics in natural rights encounters difficulty in justifying these rights.

for BCA that is grounded in pure welfarism will fail to convince many thoughtful moral agents.[8]

Benefit–cost moral theory

Donald Hubin (1994) sums it up by asking us to consider benefit–cost moral theory: the theory that right action is whatever maximizes the excess of benefits over costs, as economists understand the terms benefits and costs. It is hard to imagine a single supporter of such a moral theory, among philosophers or the public at large. Instead, we would find unanimity that such a moral theory is inadequate, and an enormous diversity of reasons as to exactly why. As Hubin speculates, most people probably believe that the recommendations of BCA are defeasible on any number of grounds.

Pluralism

A broad acceptance seems to be emerging among philosophers that the contest among ethical theories is likely to remain inconclusive (Williams, 1985). While each contending theory has powerful appeal, each is incomplete in some important way, each remains vulnerable to some serious avenue(s) of criticism, and it seems unlikely that any one will defeat the others decisively. Also, each is inconsistent with the others in important ways, so that a coherent synthesis is unlikely.

Among those who seek ethical grounding for policy prescriptions, two kinds of pluralism have emerged. The more traditional kind seeks to cultivate an intellectual environment in which people who hold resolutely to different foundational ethics can nevertheless find agreement on particular real-world policy resolutions (Williams, 1985). Agreement might be reached, for example, that real resources should be expended to protect natural environments, among people who would give quite different reasons as to why that should be so. The task of the thoughtful moral agent in the policy arena is, then, to find heuristics – rules for action – that can command broad agreement.

The second kind of pluralism imagines thoughtful people, exposed to and familiar with several foundational ethics, each calling upon different ethical traditions to answer different kinds of questions in their own lives (Rorty, 1992). To this way of thinking, if the search for the single true, complete, and internally consistent ethical theory is bound to be fruitless, exclusive allegiance to any particular moral theory is hardly a virtue; and it becomes coherent to argue that

---

[8] For completeness, it should be noted that I have defined the foundational ethics in strictly anthropocentric terms. In dealing with environmental policy issues, one also encounters: deep ecologists, for whom ultimate intrinsic value lies in naturalness; consequentialists, who are concerned with the welfare of all sentient beings; and deontologists, who argue that nature and/or many of its constituents enjoy rights that human beings are bound to respect. It safely can be asserted that the economist's anthropocentric welfarist justification of BCA makes little headway with any of these people.

some questions in life are best resolved by reference to moral imperatives, some as matters of respect for rights, while for the remainder it is reasonable to go about maximizing value, perhaps even focusing on consequences and evaluating them in terms of their impact on the level of preference satisfaction.

Pluralism and discourse

Taking seriously both kinds of pluralism encourages us to think of the policy process as inherently discursive – an open discussion among citizens searching for heuristics we can agree upon – and to accept that these rules for action are likely to incorporate insights from various moral theories.

Benefits and Costs are Morally Considerable

The failure of benefit–cost moral theory is hardly an argument that benefit–cost considerations are morally irrelevant. Hubin offers the analogy of democratic moral theory: right action is whatever commands a plurality of the eligible votes. This too is a thoroughly unacceptable moral theory. Nevertheless, democratic institutions flourish in a wide variety of circumstances, and their justification is by no means entirely pragmatic; good reasons can be found for a society taking seriously the wishes of its citizens expressed through the ballot. So, the gross inadequacy of democratic moral theory serves to justify not the abandonment of democratic procedures but nesting them within a (written or unwritten) framework of constitutional restraints, and all of this embedded in a public life where moral and ethical issues are discussed openly and vigorously.

The claim that an inadequate moral theory might nevertheless provide some principles for institutional design is entirely consistent with the standard justification of pluralism: in a world where the unique true moral foundation for public life is bound to remain elusive, public institutions should be crafted so as to make good use of insights from a variety of ethical traditions. The analogy with democratic moral theory hints at the possibility of a systematic role for BCA in public life, despite the obvious implausibility of BC moral theory.

Preference satisfaction matters

Hubin (1994) asks us to imagine a plausible moral theory in which the level of satisfaction of individual preferences counts for nothing at all. It turns out, he writes, that one cannot imagine such a theory. This claim may seem bold at first blush, but it is actually a rather modest claim because, while insisting that preference satisfaction matters, it would nevertheless permit preference satisfaction to be trumped quite readily by concerns thought morally more important. A plausible moral theory that gives no weight to preference satisfaction even in the absence of overriding concerns is truly unimaginable.

Randall (1991) and Randall and Farmer (1995) have considered the two ethical theories that contend for the allegiance of mainstream economists –

consequentialism and contractarianism – and the major alternative, Kantianism. They show that, while each of these ethical theories has different ways of taking preference satisfaction into consideration, each of them does consider preference satisfaction in some way. For *consequentialists*, one obvious way to evaluate consequences is to ask whether people prefer them; one suspects it is the default way, given that consequentialists have invested so much intellectual effort in identifying particular circumstances in which preference is an unreliable or inappropriate indicator of the goodness of consequences. It is natural, too, for *contractarians* to recommend the protection of life, liberty, and property with secure rights, in order to liberate the individual for the pursuit of happiness. Despite this respect for individual preferences, contractarians, of course, are uncomfortable with utilitarian notions of social welfare. Nevertheless, Farmer (1991) has shown that, in a world where Pareto-compensation is impeded by high transactions costs, contractarians might rationally agree to be governed by a default potential Pareto-improvement rule, provided that they are free to depart from it when more important concerns arise (the need to protect things dear to oneself, or the need to bribe aggrieved parties to remain in the contract). In the original *Kantian* scheme of things, claims based on considerations of happiness were morally subordinate to claims derived from universal moral imperatives. That, however, is less devastating to preference than some contemporary Kantians have made it seem. First, Kant himself was clear that prudential concerns matter morally.[9] Second, there may well be a broad domain of human concerns within which preference satisfaction may be pursued without violating moral strictures; and a thoughtful Kantian would concur that, within that domain, more preference satisfaction is better than less.

So the issue is not whether preference satisfaction is morally considerable; in each of these ethical theories, it is. Instead, the contest is about what sorts of considerations might trump preference satisfaction, and in what ways. What else, beyond preference satisfaction, might society want to consider, and in what manner might we want to take account of those things?

Bringing Benefits and Costs to Bear on Public Decisions

Benefit–cost analysis to inform decisions

Since preference satisfaction is a consideration under any plausible moral theory, an account of benefits and costs might be used routinely as a component of some more comprehensive set of evidence, accounts, and moral claims to inform the decision process. The notion that benefits and costs cannot always be decisive in public policy, but should nevertheless play some role, is congenial to many economists (Arrow et al., 1996: 221). However, it leaves unanswered the

---

[9] "To secure one's own happiness is at least indirectly a duty, for discontent with one's own condition under pressure from many cares and amid unsatisfied wants could easily become a great temptation to transgress duty" (Kant, 1991: 66).

question of exactly what role. Are there particular situations and circumstances in which an account of preference satisfaction should be ignored entirely, and others in which it should be decisive? How should an account of preference satisfaction be weighted relative to other kinds of information? Can the answers to these questions be principled, or must they always be circumstantial? Without more structure on the policy process itself, might not the whole business of weighing different kinds of information case-by-case bog down?

Incorporating benefit–cost considerations into a set of heuristics

The notion of a discursive policy process seeking heuristics has great appeal as a way to economize on discourse and direct it to where it can be of greatest value; that is, more toward finding rules for action and less toward debating the details of particular decisions. Might a BC decision rule be part of a coherent set of heuristics?

Routine use of BCA in public decisions would honor preferences, and guard against waste and rent-seeking. Objections to a policy role for BCA do not deny the worthiness of these considerations; rather, objections are directed toward the endowment-weighting of preference signals that is inherent in BCA, the incompleteness of welfarism as a moral theory, and the need for safeguards to ensure respect for rights that other people and perhaps other entities might reasonably be believed to hold, obedience to the duties that arise from universal moral principles or could reasonably be derived therefrom, and respect for important intrinsic values.

A benefit–cost decision rule subject to constraints

An appealing way of coming to terms with the idea that preference satisfaction counts for something in any plausible moral theory, but cannot count for everything, is to endorse a benefit–cost decision rule for those issues where no overriding moral concerns are threatened. Benefits and costs could then be decisive within some broad domain, while that domain is itself bounded by constraints reflecting rights that ought to be respected and moral imperatives that ought to be obeyed. This would implement the commonsense notion that preference satisfaction is perfectly fine so long as it doesn't threaten any concerns that are more important.

To free individuals for the pursuit of happiness, constraints securing some well-defined set of human rights seem essential. If the beneficence of reasonably free markets is to be enjoyed, secure property rights are also necessary. People acting together to govern themselves also need to establish a framework of laws, statutes, regulations, and policies, to legitimize and also to limit the role of activist government. The constitution was designed with exactly these concerns in mind.

Don't do anything disgusting

These familiar constitutional protections for life, liberty, and property take care of the major concerns of 18th-century contractarianism but, in the 21st century, we cannot stop there. The moral intuitions of the citizenry, informed by insights from the various competing moral theories, demand additional constraints. The general form of these constraints might be framed as "Don't do anything disgusting."

The basic idea is that a pluralist society would agree to be bound by a general-form constraint to eschew actions that violate obvious limits on decent public policy. This kind of constraint is in principle broad enough to take seriously the objections to unrestrained pursuit of preference satisfaction that might be made from a wide range of coherent philosophical perspectives. Examples of such constraints might include: don't violate the rights that other people and perhaps other entities might reasonably be believed to hold; be obedient to the duties that arise from universal moral principles, or could reasonably be derived therefrom; and, don't sacrifice important intrinsic values in the service of mere instrumental ends. In each of these cases, the domain within which pursuit of preference satisfaction is permitted would be bounded by nonutilitarian constraints.

The "don't do anything disgusting" constraint is congenial to those who accept the premises of ethical pluralism. Whereas a utilitarian might object that this constraint sounds fine in principle but rather empty in practice ("OK, then, tell me exactly what kinds of things are disgusting, and why."), a pluralist might respond that defining what sorts of actions are disgusting and should therefore be ruled out by constraint is exactly the right task for public discourse. Again, the pluralist sees reason to hope and expect that reasonable people can agree on particular constraints, even as they justify those constraints in quite different ways.

An Application: Conservation Policy

Consider a set of policy issues familiar to environmental economists: the protection of habitats, species, and ecosystems. A society could decide on these kinds of issues on the basis of benefits and costs, but subject to a conservation constraint. A safe minimum standard (SMS) of conservation has been suggested by a variety of authors: harvest, habitat destruction, and so one, must be restricted in order to leave a sufficient stock of the renewable resource to ensure its survival. To defend this approach, it is necessary to address some standard objections to the default BC decision rule in the environmental policy context, and to justify an abrupt policy intervention (Farmer and Randall, 1998) when the SMS constraint is reached.

Justifying BCA

In the environmental policy context, objections have been raised to endowment-weighted values, discounting future benefits and costs, and the treatment of uncertainty in BCA.

Endowment-weighted values.   Given that willingness to pay (WTP) is likely to be increasing in income and wealth, the preferences of the well-off are more heavily weighted in benefit estimation, and "well-off" must be understood broadly to include those favored by the existing pattern of rights and privileges as well as those with relatively expansive income and wealth. Given that the poor sell cheaply, costs to the less well-off are less heavily weighted. One or another manifestation of this relationship shows up in market prices, demands and supplies, and WTP and willingness to accept (WTA). To many utilitarians, the justification is by no means clear for endowment-weighting individual preferences before aggregation.

Economists recognize that endowment weighting of preferences produces benefit–cost accounts that tend to favor the status quo, which is clearly a problem.[10] Endowment-weighted valuation is, however, not without advantages. First, significant improvements in data quality arise from insisting that actions speak louder than words, and expressions of WTP and WTA speak louder than mere expressions of caring. Second, endowment-weighted valuation brings certain important classes of nonutilitarians into a pluralist consensus in support of BCA. Contractarians, of course, insist that WTP and WTA are exactly the right measures of individual welfare change. People reasoning from a variety of philosophical perspectives gain some comfort from the consideration that WTA is not strictly bound by endowments, so that the poor may have high WTA for things that they really hate to lose. It is very difficult for a proposal that would leave its uncompensated losers inconsolably worse-off to pass a PPI-based BCA test. This contractarian virtue is also likely to have some appeal for Kantians, who would recognize a duty not to impose great harm on others.

Under the PPI criterion, infinite WTA on the part of a single individual would be sufficient to forestall an otherwise net-beneficial project. This raises some serious questions: What does it mean for an individual to have an infinite WTA? And what influence should this individual position have on the collective decision? While economists seek to explain infinite WTA in terms of lexicographic preferences, I think something quite different is usually at work. Infinite WTA is typically, I think, an announcement that the individual draws upon some nonutilitarian moral tradition to address the particular kind of issue at hand. The PPI criterion – which often is criticized for failure to respect individuals sufficiently – perhaps respects individuals with infinite WTA too much. It grants an individual with infinite WTA veto power over proposed change, and it distributes this veto power more liberally than does the contractarian Pareto-safety rule (which grants veto power only to rights-holders). In a pluralistic discourse to discover an agreed-upon set of constraints on public decisions, an individual with strong nonutilitarian objections to a proposal (which is what infinite WTA most likely indicates) would be taken seriously but would not necessarily be

---

[10]   Economists differ as to how seriously they take this problem. Casual observation suggests that most proceed with fairly standard approaches to benefit–cost analysis, paying relatively little attention to this concern. A few, including Bromley (1997), go so far as to claim that it undermines the legitimacy of welfare economics – and, of course, benefit–cost analysis – as a public decision aid. Also, the debate is engaged mostly at the conceptual level. I am not aware of any systematic studies of the magnitude of distortions induced by this phenomenon.

decisive. All of this means that the PPI, by taking infinite WTA at face value, would offer quite radical protections to dissenters. In practice, BC analysts tend to adopt rules censoring or truncating extreme observations of WTP and WTA, which grants more influence, but not decisive power, to individuals with strong objections to proposed change.[11]

Discounting.   The discounting of future benefits and costs is a practice introduced from financial analysis in order to account for the productivity of capital. In recent years, some environmental economists have been swayed by critics who worry that discounting implies that the concerns of the future (perhaps only a few decades hence) count only trivially in the calculations of the present. Thus we have the discounting paradox: we must discount, it is claimed, in order to avoid damaging the future by making wasteful commitments of capital to unproductive projects; and we must not discount, it is also claimed, in order to avoid trivializing future demands for present conservation.

The paradox can be resolved in the following way. If the problem is simply to determine the rate of consumption from an endowment, a society with a positive discount rate will choose a consumption path relatively high at the outset and declining over time (Page, 1977). If capital is productive and the young need to borrow it in order to produce efficiently, equilibrium interest rates will be positive and a policy of repressing the interest rate (undertaken, one imagines, in order to protect the future) will actually depress future welfare (Farmer and Randall, 1997).[12] That is, in a cake-eating economy, discounting is destructive of future welfare, but in a productive economy it is not. A reasonable assumption is that conservation crises are most likely to be particular rather than general – involving particular resources in particular places. If this is true, policy interventions such as the SMS, which address directly crucial natural resources, provide a more appropriate response to conservation crises involving essential natural resources than would discount rate repression.

Risk, uncertainty, and gross ignorance.   Risk refers to situations in which each possible action has an array of possible outcomes, with probabilities assigned to each. A risk-neutral decision-maker will choose the action that has the highest expected value of outcomes. Risk-averse decision-makers might place more weight on avoiding unfavorable outcomes, while paying less attention to upside possibilities. Modern literature focuses on risk-management strategies, including contingent claims markets and insurance contracts. Applications to BCA (see Graham, 1981; Freeman, 1991; Meier and Randall, 1991) have shown how the

---

[11]   There is, of course, the concern that taking infinite WTA seriously would open the door to self-serving strategies. Strategic false claims of infinite WTA could deny others substantial benefits. I simply do not address this concern here: difficulties in implementing BCA are not the topic of this essay.

[12]   Farmer and Randall differ explicitly with Arrow et al. (1996) regarding the justification for discounting. We emphasize the productivity of capital and the need of the young to borrow it, whereas Arrow et al. motivate discounting by appeal to time preference.

valid conceptual measures of benefits and costs depend on assumptions concerning the completeness of contingent claims markets and the availability of fair insurance. It has been argued convincingly, I believe, that organizations with large and diverse portfolios of projects are, for that reason, efficient self-insurers, and that government is surely such an organization; therefore, government is, or ought to be, effectively risk-neutral (Arrow and Lind, 1970).

Uncertainty differs from risk, according to the familiar Knightian distinction, in that it is not possible to assign objective probabilities to possible outcomes. The Bayesian tradition suggests starting with subjective probabilities, and updating them as new information emerges.

It is often claimed that these approaches to risk and uncertainty are thoroughly unsatisfactory for public decisions about projects and policies that affect, for example, ecological sustainability. The contingent claims and insurance markets approach to risk was designed to deal with private financial risks, whereas inability to know in advance the environmental outcomes of policy decisions is quite a different conceptual problem. Rather than risk, where outcomes can be defined and probabilities assigned, or uncertainty, where outcomes are defined but objective probabilities are unavailable, advance knowledge of environmental outcomes often approaches gross ignorance: we cannot even define in advance the array of possible outcomes (Dovers, 1995). Where catastrophe (by definition, uninsurable) is thought to be one of the possibilities, approaches based on insurance theory lack credibility.

In response to concerns about risk, uncertainty, and gross ignorance, economists have offered utilitarian extensions of the BC accounts: option value (Weisbrod, 1964), quasi-option value (Arrow and Fisher, 1974), and existence value (Krutilla, 1967; Randall and Stoll, 1983). To many modern environmental economists, these categories of value are straightforward (at least in principle): they just complete the BC accounts, which began with prices multiplied by quantity changes, and gradually expanded (in recognition of the implications of the concept of preference) to include economic surplus and nonmarket use value (Freeman, 1993).

The extended BCA approach has gained a limited degree of acceptance in policy circles, but skeptics remain to the right and to the left. Supporters of business-as-usual complain that these amendments, which move BCA ever further from the discipline of the marketplace, are little more than "fudge factors" to insure that the BCA generates the environmentally correct result.[13] Environmentalists, however, are concerned that, well-meant as these amendments are, they fail dismally to capture the enormity of the risks involved and the depth of human ignorance about environmental systems.

The bottom line on BCA. Despite some persistent criticisms of BCA in environmental contexts, Hubin (1994) argues that BCA based on the PPI provides a more

[13] Ironically, perhaps, Sagoff (1996) makes similar complaints, not to keep BCA safe for market values, but to bolster his charge that extended BCA is incoherent, being a futile attempt to make utilitarian sense of intrinsic values that are inherently nonutilitarian.

plausible account of value than do its practicable utilitarian competitors, and that PPIs are at least correlated with the good of society. It makes sense, then, for citizens to take benefits and costs seriously. But legitimate concerns remain: endowment-weighted valuation provides inadequate assurance that the concerns of the poor receive attention; and BCA may be insufficiently alert to the possibility of conservation crises arising from ignorance about the workings of environmental systems and failure to foresee the future. These concerns are best addressed, I argue, by constraints on BC business-as-usual adopted for good reason. The general concerns arising from endowment-weighted valuation can be addressed by safety-net guarantees, and the specific environmental concerns by guarantees of environmental justice. The concerns about conservation crises may be addressed by SMS constraints.

Justifying the SMS

The safe minimum standard of conservation was proposed by Ciriacy-Wantrup (1968) and defended by Bishop (1978) as a rational response to uncertainty about the workings of environmental systems. Given the plausibility of carelessly exploiting a resource beyond the limits of its resilience, society should precommit to preserving a sufficient stock of the renewable resource to ensure its survival.

Utilitarian economists raised objections. First, in order to adopt voluntarily an SMS constraint, a rational utilitarian would need to have sharply discontinuous preferences. Second, Bishop's (1978) attempt to show that a risk-averse utilitarian rationally would adopt an SMS constraint – formally, the SMS is the maximin solution – failed. Writing with Ready, Bishop (1991) conceded that game theory did not support his earlier attempt at a utilitarian justification of a discrete interruption of business-as-usual when the SMS constraint was reached.

But these objections need not detain us here, because we have already conceded that an adequate resolution of the moral issues involved cannot be concerned only with preference satisfaction. Farmer and Randall (1998), arguing from existential ethical pluralism, justify the SMS as a decision heuristic adopted for good reason: a sharp break from business-as-usual, that – given the fear of possible disastrous consequences from anthropogenic modification of environmental systems about which we know so little[14] – could earn the allegiance of people operating from quite different ethical foundations, and therefore having quite different reasons for signing on. Randall (1991) and Randall and Farmer (1995) elaborate these reasons. Briefly, it is not hard to find reasons why contractarians, Kantians, and adherents of various nonanthropocentric philosophies might endorse the SMS; utilitarians are more problematic in this respect. Nevertheless, utilitarians might adopt the SMS as a binding constraint (Elster, 1979) against some impulsive and momentarily profitable act that, it is reasonably

---

[14] Taken seriously, the fear of disastrous consequences, even if quite improbable, makes a mockery of the idea that BCA, even with just the right amount of tweaking to account for uncertainty and nonuse values, can get the right answer.

sure, will be regretted eventually. While Elster's argument was not predicated on uncertainty, rational utilitarian precommitment could surely also be a coherent response to uncertainty – we would precommit not to take inordinate environmental risks in pursuit of immediate gratification.

We should be warned, however, against premature claims of consensus. The early SMS proponents, Ciriacy-Wantrup and Bishop, recognized that a utilitarian SMS could not call for unlimited sacrifice of welfare to meet conservation objectives; so, they proposed that society should be released from the conservation commitment if the costs of meeting it proved intolerably high. Randall and Farmer (1995) show that, while a case can be made for a pluralistic consensus for adoption of the SMS, controversy may reemerge concerning the magnitude of the cost that would be intolerable, especially in the case where the natural resource under threat is not essential to human welfare. Rolfe (1995) has argued that, if utilitarian thinking is used to justify the SMS, it should also be applied to determining the magnitude of the intolerable cost that would justify abandoning the SMS. By the time Rolfe is done, the utilitarian SMS amounts to little more than the extended BCA approach: a warning flag raised in information-poor situations to remind the analysts to bend over backwards to give uncertainty and nonuse values their due. In contrast, Kantians and contractarians may well, for their own good reasons, insist that the cost sufficient to justify abandoning the SMS be much larger. The work of the discursive policy process remains unfinished.

Implications for Doing Benefit–Cost Analysis

The project of perfecting BCA is doomed: BCA cannot be perfected. This would be quite terrible if benefit–cost moral theory was the one true moral theory, or if society had delegated without review all of its decisions to benefit–cost technocrats. But neither of these circumstances is actual or likely. Ethical pluralism will persist and benefit–cost moral theory is not even a serious candidate; and the pendulum began to swing away from technocracy long before the progressive dream of scientific government had been converted to reality. The issue is not how to perfect BCA, but how to enjoy the services it can provide for us – a reasonably good account of preference satisfaction, itself one valid moral concern among others – without according it more influence than it deserves. We could use BCA to inform decisions rather than to decide issues. Better yet, we could accord substantial influence to BCA, within a domain where preference satisfaction carries a good deal of weight, but bounded by various constraints derived from perhaps different ethical perspectives and adopted for good reason.

While the above argument endorses PPI-based welfare change measurement and BCA, not all accounts presented to public decision-makers and labeled as BCAs are rigorously PPI-based.[15] Some still bear evidence of BCA's roots in

---

[15] Copp's complaint (fn. 4) still has currency when addressed to the set of BCAs that are actually submitted for the attention of decision-makers.

financial feasibility analysis, while in other cases the rules for doing BCA have themselves become the subject of policy, so that deviations from the PPI criterion are institutionalized. Recently in the US, industry lobbyists have taken to arguing that, while nonmarket values and passive-use values are conceptually sound and policy-relevant, they should be excluded from BC accounts because they cannot be measured reliably; nevertheless, the resulting incomplete BC accounts should be decisive.

If BC analysts wish to claim, based on arguments such as are provided in this essay, that the public has a duty to take BCA seriously, then the analysts themselves have a duty to implement the PPI valuation framework rigorously and carefully. The result would be BCAs that depart from customary practice in several ways. Less attention would be paid to market prices and demands, while more attention would be paid to public preferences for public goods and the nonmarket values those preferences imply, and to WTA as the appropriate measure of costs. We found, much earlier in this essay, that a claimed need to impose a market-like efficiency on the activities of government provides an implausible justification for taking benefits and costs seriously. Now, we find that a sounder justification for BCA entails an obligation on the part of the analyst to pay more than customary attention to preferences and less than customary attention to market outcomes.

References

Arrow, K. and Fisher, A. 1974: Environmental preservation, uncertainty, and irreversibility. *Quarterly Journal of Economics*, 55, 313–19.

Arrow, K. J. and Lind, R. C. 1970: Uncertainty and the evaluation of public investment decisions. *American Economic Review,* 60(3), 364–78.

Arrow, K. J., Cropper, M. L., Eads, G. C., Hahn, R. W., Lave, L. B., Noll, R. G., Portney, P. R., Russell, M., Schmalensee, R., Smith, V. K., and Stavins, R. N. 1996: Is there a role for benefit–cost analysis in environmental, health, and safety regulation? *Science*, 272 (12 April), 221–2.

Bishop, R. C. 1978: Endangered species and uncertainty: the economics of a safe minimum standard. *American Journal of Agricultural Economics,* 60(1), 10–18.

Bromley, D. W. 1997: Rethinking markets. *American Journal of Agricultural Economics*, 79, 1383–93.

Ciriacy-Wantrup, S. V. 1968: *Resource Conservation: Economics and Policies*, 3rd edn., Berkeley, CA: University of California, Division of Agricultural Science.

Coleman, J. 1987: Competition and cooperation. *Ethics*, 98, 76–90.

Copp, D. 1985: Morality, reason, and management science: the rationale of cost–benefit analysis. In E. Paul, J. Paul, and F. Miller (eds.), *Ethics and Economics*. Oxford: Blackwell, 128–51.

Dovers, S. 1995: A framework for scaling and framing policy problems in sustainability. *Ecological Economics*, 12, 93–106.

Elster, J. 1979: *Ulysses and the Sirens: Studies in Rationality and Irrationality*. New York: Cambridge University Press.

Farmer, M. C. 1991: A unanimous consent solution to the supply of public goods: getting PPI rules from a PI process. *American Journal of Agricultural Economics*, 73, 1551.

Farmer, M. C. and Randall, A. 1997: Policies for sustainability: lessons from an overlapping generations model. *Land Economics*, 73, 608–22.

Farmer, M. C. and Randall, A. 1998: The rationality of a safe minimum standard. *Land Economics*, 74(3), 287–302.

Freeman, A. M. 1991: Welfare measurement and the benefit–cost analysis of projects affecting risk. *Southern Economic Journal*, 58, 65–76.

Freeman, A. M. 1993: *The Measurement of Environmental and Resource Values: Theory and Methods*. Washington, DC: Resources for the Future.

Graham, D. A. 1981: Cost–benefit analysis under uncertainty. *American Economic Review*, 71(4), 715–25.

Hubin, D. C. 1994: The moral justification of benefit/cost analysis. *Economics and Philosophy*, 10, 169–94.

Kagan, S. 1998: *Normative Ethics*. Boulder, CO: Westview Press.

Kant, I. 1991: *Philosophical Writings*, ed. E. Behler. New York: Continuum.

Krutilla, J. 1967: Conservation reconsidered. *American Economic Review*, 57(4), 777–86.

MacIntyre, A. 1979: Utilitarianism and cost–benefit analysis: an essay on the relevance of moral philosophy to bureaucratic theory. In T. Beauchamp and N. Bowie (eds.), *Ethical Theory and Business*. Englewood Cliffs, NJ: Prentice-Hall, 266–76.

Meier, C. E. and Randall, A. 1991: Use value under uncertainty: is there a "correct" measure? *Land Economics*, 67(4), 379–89.

Page, T. 1977: *Conservation and Economic Efficiency*. Baltimore, MD: Johns Hopkins University Press.

Randall, A. 1991: The economic value of biodiversity. *Ambio – A Journal of the Human Environment*, 20(2), 64–8.

Randall, A. 1999: Taking benefits and costs seriously. In H. Folmer and T. Tietenberg (eds.), *The International Yearbook of Environmental and Resource Economics 1999/2000*. Cheltenham, UK: Edward Elgar, 250–72.

Randall, A. and Farmer, M. C. 1995: Benefits, costs, and the safe minimum standard of conservation. In D. W. Bromley (ed.), *Handbook of Environmental Economics*. Cambridge, MA: Blackwell, 26–44.

Randall, A. and Stoll, J. 1983: Existence value in a total value framework. In R. D. Rowe and L. G. Chestnut (eds.), *Managing Air Quality and Scenic Resources at National Parks and Wilderness Areas*. Boulder, CO: Westview Press, 265–74.

Rawls, J. 1971: *A Theory of Justice*. Cambridge, MA: Harvard University Press.

Ready, R. C. and Bishop, R. C. 1991. Endangered species and the safe minimum standard. *American Journal of Agricultural Economics*, 73(2), 309–12.

Rolfe, J. 1995: Ulysses revisited – a closer look at the safe minimum standard of conservation. *Australian Journal of Agricultural Economics*, 39, 55–70.

Rorty, A. O. 1992: The advantages of moral diversity. In E. Frankel, F. D. Miller, Jr., and J. Paul (eds.), *The Good Life and the Human Good*. New York: Cambridge University Press.

Sagoff, M. 1981: Economic theory and environmental law. *Michigan Law Review*, 79, 1393–419.

Sagoff, M. 1996: On the value of endangered and other species. *Environmental Management*, 20(6), 897–911.

Sen, A. 1989: *On Ethics and Economics*. New York: Blackwell.

Vallentyne, P. 1987: The teleological/deontological distinction. *Journal of Value Inquiry*, 21, 21–32.

Weisbrod, B. 1964: Collective-consumption services of individual-consumption goods. *Quarterly Journal of Economics*, 78(3), 471–7.

Williams, B. 1985: *Ethics and the Limits of Philosophy*. Cambridge, MA: Harvard University Press.

5

Environmental Policy as a Process of Reasonable Valuing

Juha Hiedanpää and Daniel W. Bromley

Our task in this chapter is to argue that the standard economic approach to environmental policy is often ignored in the policy process because that approach does not fit with the fundamental character of collective action (policy) in modern democratic states. This conceptual disjunction means that economists often advance analytical frameworks (concepts, language, and methods) for assessing environmental problems that are at serious odds with the way in which ordinary citizens, government decision-makers, and politicians tend to frame environmental problems. When the policy prescriptions from economists are then ignored or denounced, economists will often express befuddlement that their "optimal" solutions were rejected. These circumstances then lead some economists to consider the policy process as irrational because it does not result in means and outcomes advocated by economists. It will be claimed that public policy is irrationally dominated by politics rather than by the self-evident rationality of science. This position then becomes the basis for often-strong advice to politicians about how environmental policy *ought to* be formulated and evaluated (Palmer, Oates, and Portney, 1995; Arrow et al., 1996).

This disconnection – often, open hostility – between policy-makers and economists is, to us, undesirable because economists have valuable and important insights and analytic approaches that could be useful to the solution of vexing environmental problems. Yet these favorable prospects are often squandered because economists persist in their belief that only economics – regarded as the science of choice – can bring rationality to the otherwise "muddled" realm of collective action. This need for the rigor and discipline of economic thought is regarded as necessary and desirable because the realm of politics – in the eyes of many economists – is dominated by selfish interests, free riders, and those who seek to get something they want at the expense of others. Aren't politicians short-sighted, misinformed, and under the thrall of special pleaders? Many economists believe this to be true.

This self-imposed mandate to impose economic rationality into the political process seems to be quite pronounced in environmental economics. For example, when economists undertake analyses and evaluations of social programs concerning, say, education, health care, old-age programs, and national defense, much of the focus is on cost-effectiveness, target efficiency, and incentive compatibility. Yet when new environmental policies are under consideration, a standard assertion by environmental economists is that such policies *ought to* be undertaken only if it can be shown that the benefits of these particular programs are in excess of the estimated costs.[1]

Let us leave aside from present consideration the value-laden assertion about how environmental policy *ought to* be formulated and evaluated. The more interesting question concerns *why* environmental economists feel authorized to insist that environmental policy must be considered through the filter of benefit–cost analysis, while other policy decisions – those concerning education, old-age programs, and health – are spared this requirement. These other social programs and policies certainly entail large levels of public spending, and they are scrutinized for their financial implications. And, of course, evidence of cost-effectiveness is a necessary condition for most everything done in the public sector. But it seems that only environmental policy is held to a standard of proof in which the benefits of environmental policies must be calculated and monetized by evidence of the citizenry's willingness to pay. These benefits must then be found to be greater than the expected costs of the proposed new policies. Lacking that proof, the proposed environmental policies are denounced as irrational, wasteful, and not in the public interest (Palmer, Oates, and Portney, 1995).

Is this different treatment for environmental policy based on a sense that existing environmental problems are *less* serious than other social choices, and hence corrective environmental policies must be held to a *higher standard* of economic prudence? Is there a legitimate concern that environmental policies hold greater potential for mistakes than is the case for education, old-age programs, health care, and the military? In other words, why do environmental economists insist that environmental policies pass this specific economistic muster when other social programs are not held to the same standards of proof about the level of monetized benefits?

It seems plausible that the insistence among environmental economists for clear quantitative evidence that environmental benefits exceed the costs of new environmental policies – and hence the great affinity for studies to ascertain the willingness of individuals to pay for environmental improvements (which are then too readily called the "social benefits" of environmental programs) – is grounded on one of several grand dichotomies in economics. The particular dichotomy at work here insists that there is one realm called *the economy* and then another quite distinct realm called *politics*.[2] With this dichotomy in hand, it becomes easier to imagine that military spending, public health, educational policy, old-age programs, and a variety of other social programs pertain to the *realm of politics*. On the other hand, many environmental economists think of – and therefore model

---

[1] We have recently seen an example of this view in a rather sharp debate about environmental regulations (Palmer, Oates, and Portney, 1995; Porter and van der Linde, 1995).
[2] See Samuels (1989).

– environmental policies as *politically imposed* modifications of the *economy.* More specifically, environmental economists tend to see environmental policies as nothing but regulatory interference with the *separate* and clearly "private" domain of firms and households. That is why the literature of environmental economics is full of frequent reference to market failure and government intervention in (or interference with) the market (Vatn and Bromley, 1997). That is, environmental policies are seen by economists as *regulatory* interventions into someone else's realm (the realm of autonomous firms and households), while health programs, educational programs, and labor programs are seen differently.

It seems apparent that much of environmental economics is but a branch of regulatory economics in which new policies are imposed on the economic realm from the external political realm. This view then requires proof that the alleged "market failure" (pollution, urban sprawl, global warming) is indeed serious enough to warrant political interference with the separate realm of the economy.[3] In an effort to protect the *economy* from this intrusion from *politics*, environmental economists will insist that a benefit–cost study be undertaken so that they might be able to ascertain – via proof of total willingness to pay for environmental improvements – that this pending intervention by the political realm into the economic realm is justified on economic grounds. If it is not clearly justified, then politics has no legitimate – by which economists mean "economic" – reason to intrude into the realm of the economy.

We will argue here that this approach to environmental policy is conceptually flawed, because it starts from a fictitious model of the policy process in democratic states. Notice that environmental economics brings not just analytic methods to environmental policy. It also brings a normative agenda in that it presumes to tell others how environmental policy *ought to* be formulated, evaluated, and implemented. Not surprisingly, this normative agenda insists that environmental policy ought to be seen as but another form of economic optimization. That is, the environment must be brought under the covering laws of economics: the environment must be commoditized, and choices about the environment must be considered as but a special case of individual maximization extended to the realm of collective action (Vatn and Bromley, 1994).

In contrast, we argue here that all public policy in democratic states is correctly understood as a process of *reasonable valuing.* With the concept of reasonable valuing in hand, we insist that it will be possible to show how environmental economists might enhance the possibility that our insights and methods might actually become useful to the policy process.

The Framing of Policy

Our argument begins with the hypothesis that environmental economics cannot usefully contribute to environmental policy when economists start with the

---

[3] This can be seen in very clear terms in the debate concerning environmental regulations and competitiveness. See Porter and van der Linde (1995) and Palmer, Oates, and Portney (1995).

presumed dichotomy discussed above. Our operating hypothesis is that it is specious to suppose that there is an economic realm that stands apart – and logically distinct – from the political realm.[4] It is logically impossible to carve the world "at the joints" such that one ends up with two distinct realms – one called the *economy* and the other called *politics*. Our strategy here shall be a realist approach. That is, we shall advance a *description* of the policy process as it exists in Western democracies. Notice that we do not start with a story about how environmental policy *ought to* be structured; the environmental economics literature abounds with such allegories. However, our description shall comprise the core premise of our inferences regarding the plausibility of the concept of reasonable valuing as how public policy actually works. From our description – as premise – it shall be possible to see exactly the ground on which we build our case for an alternative stance to the policy problem as formulated by many environmental economists. Our normative position is therefore conditional: given that public policy works in this particular way in the modern democratic state, environmental economics would stand a greater chance of making useful contributions to the policy process if it were undertaken in a manner that resembles what we shall here define as *reasonable valuing*.

We start with the proposition that environmental problems arise because of an emerging sense among the citizenry in a democratic state that particular environmental settings and circumstances – we might call them *environmental outcomes* – are becoming problematic. Perhaps particular wildlife or highly valued plant communities are disappearing. Perhaps coastal areas are too frequently coated with oil from tanker accidents. Perhaps green space in the shadow of cities is being covered over with asphalt and structures. Perhaps nearby forests are showing the effects of acid rain. Perhaps picturesque mountains are obscured by smog. Perhaps rural residents are advised against drinking well water laden with agricultural chemicals. On this account, existing or emerging undesirable circumstances become the galvanizing empirical ground for agitation on the part of some citizens. But, of course, observable outcomes are merely the tangible manifestations of millions of unobservable or simply unnoticed behaviors and practices, whose inevitable entailments comprise the objectionable outcomes.

If these suddenly objectionable outcomes are the plausible results of particular practices and behaviors, then it follows that initiatives directed at the modification of existing behaviors must focus on the *reasons for the results*. And while the proximate cause might well be particular behaviors and practices, the *reasons* for the results are the extant institutional arrangements (legal regimes and customary practices) that parameterize those now-perverse behaviors. Each of the above environmental outcomes has a plausible connection to particular behaviors and practices that are themselves products of a constellation of incentive structures that make those behaviors and practices seem – at the present time – *reasonable*. We see, in the status quo ante, prevailing notions of reasonable institutional arrangements and reasonable behaviors and practices – both predicated upon an earlier shared conception of reasonable outcomes.

---

[4] See Bromley (1989), Commons (1931, 1990, 1995), and Samuels (1989).

In the long sweep of history, it was certainly reasonable that one important purpose of rivers was to carry away the wastes of both cities and factories. Of course, transportation played a role in location decisions, but so did waste-disposal needs. It was reasonable, at the time, to discharge human and factory waste into rivers, lakes, and oceans. And it remained reasonable to do so until the entailments of those practices and behaviors became manifestly undesirable. New knowledge contributed to this nascent ambiguity concerning what was regarded as reasonable. Soon human fecal matter was discovered to hold serious public-health implications. Fish were soon discovered to die in oxygen-deprived waterways. Certain industrial chemicals were found in increased concentrations as one moved up the food chain. And it should not surprise us that attitudes and beliefs changed over time, such that the stink of rivers and lakes as open sewage systems, and the sight of poisoned fish, began to repulse the citizenry. Suddenly reasonable behaviors and practices – themselves informed by customary and legally sanctioned rules that seemed, at the time, reasonable – were realized to be causally related to environmental outcomes that no longer seemed as reasonable as they once had.

Once particular realized outcomes are judged to be unreasonable, or at least less desirable than hitherto, then it follows that the antecedent behaviors and practices associated with those outcomes warrant scrutiny. If those behaviors and practices are found to be plausible reasons for the undesirable results, then we see that behaviors and practices can also be seen as *reasonable or unreasonable* in terms of their entailments or implications. This recognition suggests the consideration of the idea of reasonable valuing.[5]

The foregoing account leads us to propose that the essence of reasonable valuing is the quest for *reasonable practices*. As we suggest above, public policy is animated by, and proceeds in the face of, emerging concerns and judgments about particular outcomes that now seem unwelcome (unreasonable). When these troublesome outcomes are identified and defined, they become *policy problems* that will motivate different segments of the community to seek a solution. In other words, the polity – whether it is local or national – faces the apparent necessity to modify or replace suddenly discredited "best" practices with new "best" practices.[6] We wish to emphasize that new policies are simply new *collective action in restraint and liberation of individual action*. New public policy – new collective action resulting in new rules for behaviors and practices affecting the environment – simply produces a modified realm of action (opportunity set) for individual choice. The search for *new best practices* that will once again produce *reasonable outcomes* constitutes the core of reasonable valuing.

Reasonable valuing calls attention to the ethical, economic, and legal circumstances in which people (as economic agents) are embedded when it is suddenly realized that existing best practices produce undesirable outcomes. This exposure

---

[5] Reasonable valuing is central to the economics of John R. Commons. See Atkinson (1987), Ramstad (1990), and Rutherford (1994).

[6] We say "best" practices to suggest that the *existing* practices need not be judged deficient in terms of their original entailments. The problem is simply that those entailments are now seen as problematic, and so the "best" practices producing those outcomes must be replaced by new "best" practices.

then offers potential avenues for new interactions among individuals and organized interests (say, farmers and environmentalists) that had hitherto been regarded as orthogonal. These new connections will strengthen some ties and weaken others. Feedback processes will be altered.

Reasonable valuing requires knowledge of the origins and historical significance of the newly harmful (or unwanted) practices. How and when were these practices *introduced*? How, and by whom, are these practices *connected* to unwanted environmental outcomes? How and by whom are these practices *interpreted and understood* as sources of existing environmental problems? In other words, whose purposes did the original practices and outcomes serve? Whose purposes are served by particular problem definitions? Whose purposes are served by alternative ways of defining and solving particular environmental problems?

Although reasonable valuing is clearly grounded on existing social practices and customs, notice that it is geared to the future. Reasonable valuing is concerned with selecting the best practices from a set of existing (and possibly) *intertwined* practices. The idea of best practices must be understood as dependent upon current perspective and the embeddedness of those practices and their advocates. The reason for seeking a solution to the newly identified environmental problem is itself the logical extension of a continuing historical lineage of conflicts and their eventual resolution under the covering law of "reasonableness."

Different conceptions of the good, reflections about means and ends, intentions, purposes, beliefs, and desires, suggest that there are a number of very different yet plausible futures. The plausibility of reasonable valuing rests on the necessity to consider these varied futures, and to evaluate not only the problematic present practices but also to illuminate the conditions of potential and possible futures. By considering present practices, economic networks and opportunities, and social relations, a *reasonable* policy process seeks to increase the capacity for responsiveness to the unfolding future. According to Commons:

> Reasonableness is a matter of judgment as well as justice, since it looks to the future effects of present acts, while justice, in itself, looks only to the past as justifying the claims of the present. (1990: 826)

The compelling logic of reasonable valuing rests on the lack of a coherent alternative. In that sense, one must see reasonable valuing as a definition – a description – of the realized outcome of a process of searching for what, under the circumstances, seems reasonable to do. Is that not, after all, the essence of the human condition? We submit that individuals and collectives do not choose what they want. Rather, they choose those things for which, at the moment, the best reasons can be mobilized. If this were not defined as "reasonable" then one would be left with the unattractive prospect of describing human history as a series of "unreasonable" actions and choices. Of course, there have indeed been unpleasant and horribly unreasonable actions in history. But it is quite improbable that the human animal could have reached its present state of evolution if all (or most) actions were patently "unreasonable."

Thinking About Reasonable Valuing

Reasonable valuing must understand the historical and institutional context of environmental disturbances and plausible solutions, it must understand the web of productive relations, it must understand reciprocity and social position in the face of transference of legal and moral rights, and it must rest on the ideal of workability.

Understanding present rules and practices

The context of environmental policy is not preordained but, rather, *emerges* with the systemic disturbances and evolving definition of the particular environmental problem that requires a solution. Following John Dewey (1939), we hold that the decision-making context encompasses the disturbed situation in which alternative courses of action are created and possible impacts evaluated. The particular emergent context depends on the specifics of the problem, the present rules and practices, and the history of the problem.

Many environmental *problems* unfortunately become environmental *conflicts* because those who seek to formulate new policies, programs, and projects appear to rule *down* – or to impose – their will on others. This imposition necessarily threatens individuals in their current behaviors and practices. Recall that existing institutional arrangements produce – they are the reason for – existing "best practices" that are suddenly found to be the plausible (proximate) cause of the newly undesirable outcomes. Most of these institutional arrangements (working rules) reveal themselves *only* in the acts they induce or compel (Commons, 1990). These working rules are themselves the outgrowth of ongoing actions and transactions within a particular social, economic, ecological, and legal environment. For instance, the force of legislative, executive, or judicial actions arises from the correlated structures and processes of sanctions, punishments, and inducements. If the law does not *act*, there are no legal working rules. In a circular and reciprocal way, rules make sense only in reference to the very regularities they are thought to bring about (Fish, 1989).

The working rules describe and prescribe which activities are permitted, which are prohibited, and which are obliged. *Prohibition* indicates which activities *cannot* be done without interference and sanctions on the part of the collective power. *Obligation* indicates which acts *must* be done on pain of sanction by the collective power. Within the realm of *permitted* actions, individuals *can* and *may* engage in practices they deem good and useful. The law is both permissive and silent.

Environmental policy – including legislation and administrative actions – may alter only the working rules that prescribe which actions shall be prohibited and obligated. That is, the law has a very limited power to influence what *actually* happens in the larger space of permitted practices. That is, environmental policy can only rank the practices according to their social or public goodness, but rarely does such policy have the ability to compel specific practices. Much of public policy concerns the structure of *normative boundaries*. How productive actions are practiced is dependent on the *moral rules* in the space of permitted actions.

The fabric of practices within the existing institutional structure is significant because particular policy measures never affect just the practice that is to be prohibited or required (obligated). Rather, interconnectedness means that changes in working rules – new policy – will echo throughout the fabric of practices (Jervis, 1999). Those effects that are intended are invariably considered to be the *results* of policy action. However, new policies perturb entire arrays of practices and may therefore give rise to *unintended effects*. Often, if the context has been properly considered, even these unintended effects could have been foreseen. Changes in the working rules may also have effects that are genuinely novel and yet unthinkable – and thus beyond the reach of deliberative policy. These surprising consequences of nonlinear interactions within the fabric of practices may be called *emergent effects*. If policy-making is considered external to the fabric of existing practices, then the insights and tools for tackling unintended consequences, and emergent effects, will be much more limited in scope than when new policy is regarded as internal to the fabric. The more complex the fabric of practices, the more likely we are to find unintended and emergent effects.

Consider the difference between emergent effects and unintended consequences. Unintended consequences are simply disturbances and surprises against the prevailing moral order, while emergent effects imply a new moral order. When programs for marketable pollution permits were introduced, some firms were reluctant to enter the market and this reluctance in turn undermined the effectiveness of the program. We see here unintended consequences; those who formulated emissions trading programs – economists, for the most part – simply could not imagine that firms would not rush immediately into this novel market. But something else happened as well. Specifically, a large number of individuals and groups mobilized against the idea of firms trading the "right" to pollute – and that opposition persists today in North–South debates over greenhouse gas emissions. If heavily polluting firms in the industrialized North can acquire pollution credits from nations in the South with large endowments of forests to sequester carbon, then opponents insist that there will be no incentive for the polluting firms to rectify their polluting behavior. We see here an emergent effect in the form of a new moral order. Suddenly, trading rights to pollute becomes a realm of dubious reasonableness. The efficiency properties of great appeal to economists are undermined by an emerging sense of moral outrage that some polluting firms can escape sanction.

The environmental policy process must be grounded on a willingness to address the key causal practices, and to assess how these practices lead to the unwanted outcomes. It is also necessary to understand the ways in which these practices are interconnected. In order to provide sensitivity and responsiveness to the context, a "map" of current practices must be drawn. This map will be helpful in illustrating how the environmental disturbances first emerged, whose purposes those practices served, and according to which parties the disturbances are interpreted as environmental problems warranting rectification.

John R. Commons (1990) regarded institutions as (the result of) collective action that constrains, liberates, and expands individual action. He further divided collective action into two types: organized and unorganized. *Unorganized collective* action concerns taboos, traditions, conventions, customs, etiquette, and routines associated with individual habits and tastes. These unorganized forms of collective action differ from one another in both space and time; traditions endure longer than individual habits and tastes. As Max Weber once noted: "Convention transforms custom into tradition" (Weber, 1968: 326). *Organized collective* action concerns, for example, the European Union, nation–states, firms, communities, associations, and assorted groups. These realms of collective action differ from one another in one specific respect: the extent to which the rules of action are articulated and formalized. Organized collective action is purposeful and bound by formal rules. Unorganized collective action is purposeful, the working principles of taboos, traditions, and custom are followed, but the working *rules* are not always clearly articulated or thoroughly specified.

The purpose of reasonable valuing is to select the best existing practice, and here it is necessary to know how and why some practices have persisted, some have evolved, and some have perished in the course of time. As Richard Bernstein (1983: 130) has noted, "all reason functions *within* traditions." Scientific habits of thought and action *are* themselves the "working rules" for finding answers to questions about problematic practices and the working rules that allegedly induce those actions. This is why multidisciplinary work is so difficult to carry out, and so important in public policy.

The nature of economic relations

While the individual is regarded as the sufficient unit of analysis in the standard economic approach to environmental policy, a more realistic conceptualization would reveal that the pertinent entities must be seen as *organized groups* of individuals. As above, *organized* collective action is the realm in which the reigning social goals and rules are identified, articulated, and adopted. The relevant entities are firms, households, government agencies, trade and labor organizations, commercial associations, nation–states, and political (environmental) movements (Commons, 1990). With much economic attention currently devoted to households and firms as realms of contracts and negotiation, this movement beyond the isolated individual is now well accepted in much of economics.

William Connolly (1999) uses the expression "the politics of becoming" to suggest that policy-making must be sensitive to the future – indeed, sensitive enough to acknowledge the emergent changes in the fabric of practices, and changes in group composition. This should happen when new practices and groups are coming into being, not when they have become rigid and embedded. When the fabric of practices and economic opportunities change, the purposes and intentions of people are under duress. Policy processes must be sensitive and responsive to what the future may bring. The politics of becoming entails the *historical sense* acting in the *present*, for potential and *possible futures*.

Connolly uses another expression as well – "the politics of collective assemblages" (1995). The politics of collective assemblages addresses the issue of how different changes in the fabric of practices might affect individuals' security of expectations, conformity, liberty, and exposure provided by different organized groups. The purpose here is to discover the key groupings: (1) those who have a special role in producing the environmental disturbances or problems; (2) those who play a significant role in keeping the system stable; and (3) those who play an important role in solving environmental problems. Thus, the politics of assemblages is a tool of the policy process in examining structural and functional features of the locality.

Reciprocity

Reasonable valuing is predicated on the notion that the new prohibitions and obligations must not be the starting point of environmental decision-making. Because reasonable valuing entails cooperation and concepts of fairness and mutuality, new policy measures will be most effective if developed in the general realm of existing permitted actions. That is, new policy should, to the extent possible, be connected in some way with those activities in which people are already engaged.

Policy conflicts arise because of a tension between *legal* and *moral* perceptions of rights and duties. This tension emerges because those who currently hold rights tend to regard their favored position as coincident with the larger public purpose, and hence duties must be imposed on those who would threaten this conflated private/public purpose. Most environmental problems – leading to agitation and perhaps wider social conflict – challenge the prevailing public purpose (and the prevailing presumptive rights structure). In contrast, social consensus implies that the practices, economic opportunities, and social relations and positions are accepted. As Connolly has observed, the presence of consensus (the absence of conflict) is a sure sign of danger, because this suggests that the policy process is not sensitive to pressures and sufferings related to the acts of resistance of unfolding practices, identities, and collective entities that are emerging within. The absence of consensus (the presence of social conflict) is not necessarily good either, because then the contours between the groups, identities, and practices are thick, exclusive, and self-centered (Connolly, 1995). This suggests that the policy process in democratic states operates at the *edges of consensus and conflict*.

A common feature between reasonable valuing and the other economic approaches to value is that people are considered as instrumentally rational (*zweckrational*) beings: by their very nature, people try to select the best possible means to attain desired and admired goals or ends-in-view. The purpose of reasonable valuing is to examine different perspectives to the best possible means (practice) for given ends.

Deliberations of ends-in-view and means are clear signs of the use of reason and rational inference, but what makes this process reasonable is that ends-in-view and means are reflected in the light of economic relations and social positions – together with actual and potential practices. Reasonableness is assessed in the face of a multiple set of values, articulated reasons, and unarticulated causes for regarding certain ends as good and certain others as evil – and certain means right and other means wrong (bad).

The social psychology of reasonable valuing is based on innovative reciprocal learning processes, not on the commands, control, and exclusion exercised by legal superiors. Reasonable valuing works actively against "passive exclusion" and is for "active inclusion." Not every player may have legal rights, but with the means of reciprocity an attempt is made to give rise to new moral insights about current rights and values. In the process of active inclusion, social ties and links are strengthened and reconstructed, which may provide more reflective conditions for social relations and moral claims. People tend to show more respect to those who belong to the same particular "we."

Workability: the only ethical ideal

The essential purpose of reasonable valuing is to select the best existing practices as a resolution for particular environmental disturbances and problems. But how do we recognize the *best* practices? Because reasonable valuing is a process that takes place at the point of a particular policy conflict (locality), there is no outside vantage point from which to assess and weigh different practices. The criteria for finding or creating the best practice are always local (Fish, 1999). Depending on existing practices, the "best" varies from one viewpoint to the other. It might be said that "where you stand depends on where you sit." The *best* is always relative to one that serves the "best" individuals and groups in question. Because people do not always know what is the best for them, it is essential that the context of the policy process is sufficiently open and transparent for all participants. This gives everyone the best possible opportunity to judge what they regard as "the best." This judging cannot be the view of people acting somewhere outside of the "locality" (Benhabib, 1992).

But how are the best existing practices selected? If policy-makers, scientists, groups, and individuals exposed to particular environmental disturbances and problems are members of the same locality, and none of them is in a sovereign position to rule down (impose) solutions to the problems, who then does the selecting? There are two types of selection that must be considered: (1) natural selection and (2) artificial selection.

Natural selection refers to the process in which "average" conditions in a setting are responsible for the selecting. Natural selection cuts down (cuts out) the traits and characteristics that do not fit the prevailing social and economic context. Those traits and characteristics may be good or bad, but they are "singled out" for elimination because they deviate from the average – from the norm. In a sense, certain traits and practices are allowed to die out because, as above, they no longer "fit" the evolving circumstances. Artificial selection, on the other hand, is not a passive or even mechanical process in which certain attributes are eliminated (or discontinued) because they do not "fit in." Rather, artificial selection is a process of purposeful action by individuals or groups with an end-in-view (and -in-mind). The pertinent actors cut out those traits and characteristics that do not fit their purposes.

Workability implies that the policy process selects the *best* out of many practices considered best. For this difficult task, Richard Rorty offers the following advice: "We shall call 'truth' and 'good' whatever is the outcome of free discussion – if we take care of political freedom, the goodness and truth will take care of themselves" (Rorty, 1989: 84). There are two key ideas here: *free* discussion and political *freedom*. Free discussion entails a shift from *argumentation* to *articulation*, and from *commanding* to *persuasion* (Rorty, 2000). Argumentation refers to rationality: the rules of action and inference are preknown. Articulation refers to reasonable. Free discussion understood as a process of articulation means that everybody is given a *voice*. Free discussion is, however, empty without ethical responsiveness and solidarity within the locality. Free discussion entails that purposes, economic opportunities, and social ties are so tightly interconnected that the *noise* within the fabric of practices turns out to be a signal – a voice with a purpose.[7] The more sensitive actors are to turning noise into purposeful signals, the freer is the discussion in given conditions. Free discussion only attempts to make these relations more tangible and articulate – and people more sensitive to them. If things go well, the consequence of articulation may be an increase in the incidence of forbearance in reciprocal relations.

We interpret political freedom in Deweyan terms, and consider it as the potential and possibility for novel actions, personal growth, social learning, and technical (artistic) development (Dewey, 1999). Taken this way, political freedom refers neither to positive or negative freedom. In the pragmatist conception of freedom, the question is not about to be *free to* or *free from*, but to have an active capacity and willingness to exercise the *freedom in* existing conditions (Commons, 1995).

Reasonable valuing – free discussion and the exercise of political freedom – could not take place without the willingness for ethical growth and instrumental learning. Ethical growth means that people learn a new way to talk about old problems and practices, or that they learn to speak about things that they did not know before. Instrumental learning, on the other hand, means that people learn to practice a new thing, or that they learn to perform an old task in a new way. Ethical growth and instrumental learning are intertwined: learning to talk about new things implies that something new has happened that was incomprehensible before, and learning to do a new thing cries out for words to understand what has happened. To paraphrase Ludwig Wittgenstein, our world ends where our language ends – and vice versa. Ethical growth and instrumental learning in reasonable valuing means that the space of possibilities expands; and in doing so it affords more potential actions for people and groups. People may exercise more freedom.

Strengthening positive feedback within the space of permitted actions is the only means to strengthen the economic and social relations between people within the given locality. By strengthening these ties, the ethical networks grow stronger; the "*we*" grows larger.

---

[7] See Hodgson (1999). His discussion concerns the necessary impurities in all economic systems.

Why Reasonable Valuing?

We suggest that the essential purpose of environmental economics is to provide knowledge about the effectiveness of different environmental policy alternatives, where *effectiveness* may well be considered in terms of *cost*-effectiveness. Because economics tends to accept the status quo, environmental economists do not have a reason to question how existing practices are ranked. By changing the rank ordering, the cost-effectiveness of new policy proposals will change. Notice that in conventional environmental economics the merits of any policy change is always judged against the status quo. Unlike these conventional approaches, reasonable valuing examines the whole fabric of actual, potential, and possible practices. For this reason, reasonable valuing applies multiple criteria for ranking practices and assessing the cost-effectiveness of policy proposals.

Standard environmental economics provides information about individual preferences concerning certain environmental goods and services. This information is considered necessary for assessing the economic value (worth) of policy proposals. In these studies, people are either asked how much they are willing to pay to experience an improvement in some aspects of their environment (willingness to pay, WTP), or they may be asked how much they believe they should be compensated in order to accept (willingness to accept, WTA) the continuation of the status quo ante. According to reasonable valuing, these are the wrong questions. Better questions are: How much disturbance will people *resist* (WTP)? How *resilient* (WTA) are their preference structures? The resilience and resistance of individual character, identity, sense of duty, righteousness, and, say, virtuousness in the face of environmental disturbances and changes are not only due to their economic imagination or actual economic relations and opportunities. Resilience and resistance also depend on moral and political values that the extant intertwined economic and social relations cultivate and nurture. Unlike received environmental economics, reasonable valuing explicitly addresses the multiple aspects of individual and group resistance and resilience in environmental planning and decision-making.

Reasonable valuing considers individuals as situated, embedded, and *embodied* actors within the evolving dynamic economic and social structures and relations of the locality. Values and preferences are not fixed, but are created and recreated while transactions are under way. Reasonable valuing focuses on the multiple ways in which practices, groups, and norms "scaffold" individual choice sets for different actions (Clark, 1997). Reasonable valuing provides better grounds for the environmental policy process than the established forms of mainstream environmental valuation, that consider individuals as rational and conscious in their actions, and independent and autonomous from other actors, practices, and norms.

Traditional environmental economics operates outside of the policy process, and the traditional policy process operates outside of the given locality. Mainstream policy ignores complexity, embeddedness, and the multilevel dynamics of locality. Mainstream economics *explains* environmental changes in mechanical terms. Reasonable valuing operates within the locality and thus offers better grounding for *understanding* the causes, reasons, and purposes behind particular environmental problems. This includes an understanding about the complex working principles of practices, groups, and norms. Only then is it possible to anticipate what Commons calls the *purposes of the future* – where the present is taking us, and why. Standard environmental economics is concerned with question of *what* and *how*. Reasonable valuing is concerned with questions of *how* and *why*.

Mainstream environmental economics does not pay attention to the complex ways in which people discuss environmental values. Reasonable valuing acknowledges that language plays an important role in the process by which individuals get entrained and attuned to certain practices, purposes, and identities. Also, reasons for certain environmental activities (values) tend to be justified by utilizing certain individual vocabularies and organizational narratives (Czarniawska, 1997). Because the mainstream is trapped in a discredited view of language as a *medium* between individual and outer world, there is a general inability to grasp the importance of language in the policy process. Both the standard policy process and environmental economics tend to deal with "real" things – problems and people – rather than with those things that stand *between* the real things. In contrast, reasonable valuing considers language as a *tool* for coping with the environment. That is, language is not a neutral medium, but constitutes an active tool in changing the patterns of interactions and relations in the particular environment (Clark, 1997). Reasonable valuing sheds light on *vocabularies* and *narratives* that constrain, liberate, and expand the ways of justifying purposes, causes, intentions, preferences, and actions.

In sum, reasonable valuing is a process of criticism and innovation. Reasonable valuing is a process in which existing conditions, and potential and possible futures, are challenged. Reasonable valuing is the practice of (constructive) criticism in search of reasonable social practice.

References

Arrow, K. J., Cropper, M. L., Eads, G. C., Hahn, R. W., Lave, L. B., Noll, R. G., Portney, P. R., Russell, M., Schmalensee, R., Smith, V. K., and Stavins R. N. 1996: Is there a role for benefit–cost analysis in environmental, health, and safety regulation? *Science*, 272 (12 April), 221–2.

Atkinson, G. 1987: Instrumentalism and economic policy: the quest for reasonable value. *Journal of Economic Issues*, 21(1), 189–202.

Benhabib, S. 1992: *Situating the Self: Gender, Community, and Postmodernism in Contemporary Ethics.* New York: Routledge.

Bernstein, R. 1983: *Beyond Objectivism and Relativism: Science, Hermeneutic, and Praxis.* Philadelphia: University of Pennsylvania Press.

Bromley, D. W. 1989: *Economic Interests and Institutions: The Conceptual Foundations of Public Policy.* Oxford: Blackwell.

Clark, A. 1997: *Being There: Putting Mind, Body, and the World Together Again.* Cambridge, MA: The MIT Press.

Commons, J. R. 1931: Institutional economics. *American Economic Review*, 21, 648–57.

Commons, J. R. 1990 [1934]: *Institutional Economics: Its Place in Political Economy.* New Brunswick, NJ: Transaction Publishers.

Commons, J. R. 1995 [1924]: *Legal Foundations of Capitalism.* New Brunswick, NJ: Transaction Publishers.

Connolly, W. E. 1995: *The Ethos of Pluralization.* Minneapolis: University of Minnesota Press.

Connolly, W. E. 1999: Suffering, justice, and the politics of becoming. In D. Campbell and M. J. Shapiro (eds.), *Moral Spaces: Rethinking Ethics and World Politics.* Minneapolis: University of Minnesota Press, 125–53.

Czarniawska, B. 1997: *Narrating the Organization: Dramas of Institutional Identity.* Chicago: University of Chicago Press.

Dewey, J. 1939: *The Theory of Valuation.* Chicago: University of Chicago Press.

Dewey, J. 1999 [1935]: *Liberalism and Social Action.* Amherst, NY: Prometheus Books.

Fish, S. 1989: *Doing What Comes Naturally: Change, Rhetoric, and the Practice of Theory in Literature and Legal Studies.* Durham, NC: Duke University Press.

Fish, S. 1999: *The Trouble with Principle.* Cambridge: Harvard University Press.

Hodgson, G. 1999: *Economics and Utopia: Why the Learning Economy is Not the End of History.* London: Routledge.

Jervis, R. 1999: *Systems Effect: The Complexity of Political and Social Life.* New York: Princeton University Press.

Palmer, K., Oates, W., and Portney, P. R. 1995: Tightening environmental standards: the benefit–cost or the no-cost paradigm? *Journal of Economic Perspectives*, 9(4), 119–32.

Porter, M. E. and van der Linde, C. 1995: Toward a new conception of the environment–competitiveness relationship. *Journal of Economic Perspectives*, 9(4), 97–118.

Ramstad, Y. 1990: The institutionalism of John R. Commons: theoretical foundations of a volitional economics. In W. J. Samuels (ed.), *Research in the History of Economic Thought and Methodology.* Boston: JAI Press.

Rorty, R. 1989: *Contingency, Irony, and Solidarity.* Cambridge: Cambridge University Press.

Rorty, R. 2000: *Philosophy and Social Hope.* New York: Penguin.

Rutherford, M. 1994: *Institutions in Economics.* Cambridge: Cambridge University Press.

Samuels, W. J. 1989: The legal–economic nexus. *George Washington Law Review*, 57(6), 1556–78.

Vatn, A. and Bromley, D. W. 1994: Choices without prices without apologies. *Journal of Environmental Economics and Management*, 26(2), 129–48.

Vatn, A. and Bromley, D. W. 1997: Externalities: a market model failure. *Environmental and Resource Economics*, 9, 135–51.

Weber, M. 1968: *Economy and Society: An Outline of Interpretive Sociology.* Berkeley: University of California Press.

PART III

Ethical Concerns and Policy Goals

CHAPTER 6

Rethinking the Choice and Performance of Environmental Policies

Jouni Paavola

[T]there can be no way of justifying the substantive assumption that all forms of altruism, solidarity and sacrifice really are ultra-subtle forms of self-interest, except by the trivializing gambit of arguing that people have concerns for others because they want to avoid being distressed by their distress. (Elster, 1983: 10)

The conventional economic approach to environmental issues goes against the common sense of many individuals. We see clear evidence of this conflict of views in the highly publicized protests against the World Trade Organization and the increasingly global market economy. The conflict is also evident in the everyday pursuits of environmental organizations, who often (but not always) reject the prescriptions of economists and sometimes pursue antithetical courses of action. Even closer to home, politicians often ignore economists' advice – witness the popularity of "command and control" measures in comparison to environmental fees and trading systems favored by economists. Finally, most of us probably feel that collective choices have – or at least *should* have – a broader basis than is offered by the conventional economic approach.

The conventional economic approach is indeed in a somewhat schizophrenic position when addressing environmental issues. On one hand, the received wisdom calls for respecting individual preferences, whatever they are. On the other hand, the conventional economic approach assumes that rational agents do not pursue goals other than their own personal utility or welfare. That is, the utility or welfare of the choosing agents is considered the only motivation that does and can inform individual preferences. This is indeed a curious way of respecting individuals' preferences – they are respected only so far as they are what the economists say they should be.

Yet it is perfectly imaginable that well-informed agents could prefer courses of action – or choice alternatives – that are not in their own immediate best interest. Bothering to go to the polls to cast one's vote, when voting can only have a marginal influence on the outcome of elections, is often cited. Contingent valuation studies also provide evidence that individuals pursue goals other than their own personal welfare (Spash and Hanley, 1995; Jorgensen et al., 1999). We also make choices in our everyday life that are not in our best immediate interests. There is nothing wrong with this: acting against one's own immediate interests in order to realize some other good is part of an age-old notion of what being honorable means.

I argue below that we should often respect individual preferences although they would not forward the immediate best interests of their holder, at least when preferences are well-informed. Correspondingly, I argue that we should pursue other goals in addition to (or instead of) social welfare when making collective choices concerning environmental policy. My argument for taking individuals' preferences instead of their welfare seriously does not mean that their welfare does not matter. When we accept that some (but not all) individuals may prefer alternatives that are not in their own immediate best interest, we endorse pluralism of values in its fullest sense. This means that individual standpoints *vis-à-vis* a particular policy choice may well be incommensurate – there may not be a way to find a solution that is optimal in the conventional sense. Yet it is possible to make collective choices. That is, collective choices follow a different logic than is implied by the conventional economic approach. When pluralism of values prevails, deliberation balances welfarist and nonwelfarist concerns in collective choices. As a result, some collective choices are informed predominantly by welfarist concerns while others are not.

I suggested above that we should usually respect *well-informed* preferences although these preferences would not advance the immediate best interests of the individual. Being well-informed does not mean an awareness of the true costs and benefits of a course of action for oneself and for the others. Being well-informed means having deliberated on the alternative courses of action and having consciously formed one's preferences over these courses of action. Conscious moral choices that go against one's own immediate best interests are well-informed and merit respect both in actual choice contexts and in the analysis of choice in economics. Uninformed preferences do not necessarily merit the same respect. As Olof Johansson-Stenman suggests in the following chapter of this volume, there are also good reasons to forward what we deem to be in the best interests of the affected individuals when preferences are uninformed. There are also some choice outcomes, such as physical harm to other humans, that we simply do not and should not accept (or make legitimately available), no matter how well informed agents' preferences are with respect to these outcomes.

We should thus not expect that one set of values – such as welfarism – is universally applicable as a guideline to policy and other choices. We should instead be sensitive to what values actually inform choices in different choice contexts. In what follows, I will elaborate and justify these arguments in greater detail, and explore their consequences for our understanding and analysis of the choice and performance of environmental policies.

Economic Analysis and Motivations for Environmental Protection

Conventional environmental economics assumes that all agents are motivated by their utility or personal welfare. For example, agents who reveal preferences for environmental protection are thought to expect and to obtain welfare gains from the maintenance or improvement of environmental quality. Moreover, environmental economics usually assumes that these welfare concerns exhaust agents' motivations for environmental protection. Monetary valuation of the environment is based on this assumption: rational agents are thought to be willing to pay at least an amount that equals the value of environmental quality in order to secure it. The monetary value of the environment (or a change in its quality) can thus be measured by determining agents' revealed or stated willing- ness to pay for it.

As a corollary of these behavioral assumptions, environmental economists often advocate setting welfare-maximizing environmental goals, and the adoption of cost-effective instruments of environmental policy. However, these prescriptions have not proved popular in practice. Policy-makers continually choose policy goals and instruments that economists do not consider welfare-maximizing, and they have failed to replace existing policy goals and instruments with alternatives that economists consider superior in welfare terms. For example, risk policies do not equate the marginal costs of saving additional lives (Huber, 1983). More- over, "command and control" measures are still more widely used than environ- mental charges and trading systems favored by economists. Environmental activists and the general public are also sometimes discontented with the way in which conventional environmental economics approaches environmental issues and informs policy debates and choices. Recent reactions against proposals to use flexible mechanisms and carbon sequestration to mitigate climate change seem to convey this discontent.

This discrepancy between policy prescriptions and policy choices does not necessarily mean that policy choices are as bad as some critics seem to suggest. Rather, conventional economic analysis has failed to acknowledge factors that characterize actual policy choices. One plausible explanation for "nonmaximizing policy choices" is that citizens may not always seek to improve their personal welfare, or even some notion of social welfare through collective choices. There are good reasons to take this possibility seriously. First, empirical evidence suggests that agents do not seek to protect the environment only because of expected welfare gains. In contingent valuation studies, respondents sometimes express strong com- mitments to environmental protection while refusing to offer willingness-to-pay estimates (Spash and Hanley, 1995; Jorgensen et al., 1999). Willingness to pay may thus not be a proper measure of preferences for environmental protection. Second, there are also persuasive philosophical and theoretical arguments – some of which will be examined below – for doing so.

It thus seems warranted to recognize a broader range of behavioral motivations than has been customary in conventional environmental economics. To make space for these motivations, including those that are not related to the welfare of the choosing agent, the narrow understanding of rationality must first be re- placed with a wider notion of rationality as intentional action that is consistent with the reasons and plans of an agent. For Elster (1983), this kind of rationality is "thin," and it also resembles Simon's (1978) notion of procedural rationality, which does not present substantive requirements for rationality. The broader notion of rationality readily accommodates the pursuit of other goals by agents in addition to their personal welfare.

The concepts of "utility" and "welfare" also need to be clarified for the purposes of this chapter. The problem with the conventional use of these terms in economics is that they are often explicitly or implicitly conflated. This confusion has historical roots. For early neoclassical economists, the concept of utility referred to both *usefulness* and *pleasure*, usefulness relating to objective needs or wants, and pleasure to subjective desires. Both aspects of utility had their analytic uses. The idea of utility as a measure of usefulness forged a strong link between utility and welfare, and enabled cardinalists to make interpersonal comparisons in a number of choice contexts (Georgescu-Roegen, 1968; Cooter and Rappoport, 1984; Sen, 1991). After all, individuals do have broadly similar objective needs for sustaining themselves – an idea forwarded today by Amartya Sen – although their subjective desires may vary greatly. The ordinalists understood "utility" as a measure of the satisfaction of desires. While this understanding retained the link between utility and welfare, it made interpersonal comparisons of utility difficult, if not impossible.

The association between utility and welfare was finally broken when Hicks and Allen (1934) redefined utility as a measure of the degree to which the agents' preferences are satisfied. This ambitious redefinition enabled any values to inform preferences. Yet the association between utility and welfare has survived in the minds of most economists: the maximization of utility is thought to imply the maximization of welfare as well (Sen, 1991). However, there is no guarantee of this outcome when utility is a measure of satisfaction of preferences, including nonwelfarist ones. Therefore, I shall abandon the broad concept of utility as a measure of preference satisfaction. For further clarity, I avoid using the term "utility" and use the term "welfare" when referring to the old core meaning of utility as usefulness or pleasure. This terminological practice helps us to distinguish between welfarist (traditional utilitarian) and nonwelfarist (nonutilitarian) motivations, and to examine their implications.

The recognition of a wider range of behavioral motivations means accepting both intrapersonal and interpersonal value pluralism. *Intrapersonal value pluralism* means that an agent may simultaneously hold different values that could inform his or her choices in a choice situation. An example is a situation in which an individual has to choose between going out with his or her friends or visiting his or her elderly mother: the first alternative may promise more pleasure, but the individual may also feel an obligation to do the latter. In other words, agents must often first *choose between values*, which choice will then serve to inform their preferences (Anderson, 1993). Kavka (1991) argues that when we admit intrapersonal value pluralism, we must attribute to individual choice all the complications indicated by impossibility theorems that we usually consider to characterize collective choice. *Interpersonal value pluralism*, in turn, means that different agents may be informed by – and act upon – different values in the same choice situation, and arrive at either similar or different choices.

Furthermore, value pluralism may refer to both the substance and the form of values, of which the latter is more consequential. For example, two self-interested and welfare-centered agents may value environmental quality differently: one considering it unimportant and the other important for his or her welfare. Still, the choices of these two agents do reveal what they consider as best enhancing their welfare. However, when agents' preferences are not based on self-centered welfare considerations, their choices no longer reflect what is best for their welfare. For example, social welfarists and other-regarding welfarists could choose so as to improve the welfare of other humans (or even nonhumans). Nonutilitarian consequentialists could assess choice alternatives with respect to the consequences they consider intrinsically valuable. A Kantian rule-following agent could in turn rule out certain alternatives because he or she thinks that choosing them would simply be wrong, no matter what the balance of good and bad consequences would be.

To restate the obvious, *choices do not and cannot reveal preferences when agents are motivated by plural values*. For example, knowing that an individual has chosen to observe a vegetarian diet does not reveal the motivations that led her to choose vegetarianism. She may have chosen vegetarianism so as to improve her own personal welfare, but there are also other possible explanations. She may have chosen vegetarianism because of concerns for the welfare of other humans or animals. Alternatively, she may have concluded that animals should not be killed because their life is intrinsically valuable. Finally, she may have considered that being a vegetarian is simply the right or virtuous thing to do. It is important to entertain the possibility of all of these motivations, because they entail different choice behavior. All of these motivations could lead one to become a vegetarian. Yet nonwelfarist agents would choose to become vegetarians under personally more unfavorable conditions than welfarists, and they would be less likely to revoke their choice if other things do not remain the same. It is also worth noting that an other-regarding welfarist could abandon vegetarianism if she could be convinced that the husbanding of animals for meat does not unreasonably reduce their welfare. The same argument is unlikely to have an influence on a nonutilitarian consequentialist or a Kantian vegetarian.

We see that preferences for environmental outcomes can thus be based on self- or other-regarding welfarist values, nonutilitarian consequentialist values, or deontological values. The preferences of a self-interested welfarist could not induce her to act or choose in ways that decrease her welfare. An agent whose preferences are shaped by other-regarding welfarist or nonwelfarist environmental concerns (which include environmental concerns based on nonutilitarian consequentialism and deontology) in turn could do so. Ethical premises capable of inducing welfare-reducing behavior do not, by any means, influence attitudes toward the environment only. Quite the contrary – nonwelfarist ethical premises are important in all institutional choices. For example, human rights are often promoted or defended because they enable broad human agency that is considered intrinsically valuable. Alternatively, respect of human rights may be considered a duty or virtue without reference to its consequences.

It is relatively straightforward to incorporate value pluralism into economic analysis of individual choice. An agent's preferences are usually understood as that ranking of choice alternatives that would maximize her welfare. The agent is also assumed to choose according to her preferences so as to maximize her welfare. When agents have nonwelfarist concerns, the connection between either preferences and choice, or choice and welfare is severed (Sen, 1973). Preferences can be understood to reflect only the welfare an agent expects to derive from various choice alternatives. In this formulation, nonwelfarist values influence choices directly – the agent sometimes chooses against her preferences. Alternatively, preferences can be understood to include also those rankings of choice alternatives that are not based on welfare comparisons – the agent sometimes prefers alternatives that decrease her welfare, and yet she chooses them. The formulation in which preferences are understood conventionally is often followed (Sen, 1973; Bromley, 1989) because it does not directly question the conventional analytic framework. However, the formulation in which preferences include nonwelfarist rankings, and in which choices do not relate to the welfare of the choosing agent, presents a more accurate and balanced view of choice when values are plural. Therefore, the latter formulation will be followed below.

Rethinking the Choice of Environmental Policies

Conventional environmental economics does not usually present an explicit theory of policy choice, although one is certainly implicit. This implicit theory holds that welfare-improving policy choices are selected because they are beneficial to welfare-centered agents. This is a "naïve" theory of institutional change (Eggertsson, 1990): it does not explain how agents' pursuit of their self-interest is translated into public policies that improve social welfare. Yet the implicit theory of policy choice is a good starting point when examining the difficulties of using the standard economic framework to explain nonmaximizing policy choices. I will gradually add complexity to the analysis, and I will demonstrate that the difficulty of explaining nonmaximizing policy choices in the conventional framework does not easily vanish. I will conclude the section by outlining a pluralist theory that explains at least some policy choices that do not improve social welfare.

Environmental economists seem to argue that policy choices that do not maximize social welfare – or that may even reduce social welfare – are quite common. For example, the post-1972 Clean Water Act Amendments have been argued to have produced a loss in social welfare (Freeman, 1990). Ackerman and Hassler (1981) have concluded that the New Source Performance Standards enacted in 1979 under the Clean Air Act of 1970 for new coal-burning power-generating plants have compromised both environmental and welfare goals. Many economists also condemn the use of "command and control" measures as wasteful of resources (Hahn, 1989; Palmer, Oates, and Portney, 1995). Let us try to explain these arguably nonmaximizing outcomes for an argument's sake.

First, we notice that it is difficult to explain these nonmaximizing collective choices from within the conventional economic framework. If agents are self- and welfare-centered, all-knowing, and capable of making welfare-maximizing choices in an idealized market, they should also be able to reach collective choices that maximize social welfare in an ideal political arena where agents enjoy equal power, transactions are costless, and collective choices are voluntarily agreed upon (assuming that entitlements have been fully defined and assigned before collective choices commence). In this kind of polity, those benefiting from a new policy would compensate adversely affected agents in order to obtain their support for a policy change. The adversely affected agents would consent to support the policy change if they were adequately compensated because they would consider adverse effects and compensation commensurable. Therefore, all collective choices reached in this ideal polity would pass an actual compensation test and be true Pareto improvements.

The ideal polity clearly does not characterize actual polities, but we can increase the degree of realism by introducing majority rule. Majority rule makes compensation unnecessary and allows the majority to benefit at the cost of the minority, in theory even to the detriment of social welfare. However, this is likely only when the majority and minority coalitions remain the same, or when collective choices are not repeated. Conversely, compensation in some form ensures cooperation when repeated collective choices are made and majority and minority coalitions do change. Namely, under these conditions agents remain uncertain as to whether they belong to the minority or the majority in the future. Therefore, agents have to mitigate the worst outcomes and seek to ensure cooperation that benefits them (Rawls, 1971; Axelrod, 1984). In "normal" polities, cooperation is ensured either by directly compensating the losers of a policy choice, or by reciprocating – letting the losers of one policy choice win in another policy choice. Notice that perfect reciprocity, because it institutes the actual compensation test across policy choices, eliminates the possibility of welfare-reducing collective choices. Of course, agents in actual polities do not exhibit perfect reciprocity. If the political system distributes power asymmetrically and provides unequal participation in collective choices, nonmaximizing policy choice can emerge. However, admitting this requires violation of the conventional economic assumptions.

It is also possible to argue that nonmaximizing collective choices are mistakes. Many collective choices are indeed complex, and made on the basis of uncertain and imperfect information. However, collective choices are no different in this respect from other choices, and could be claimed to exhibit fewer informational problems than choices made in markets by households and firms. Individuals have limited resources to seek information on alternatives in choice situations that they face infrequently. This is why "lemons," or substandard goods, change hands (Akerlof, 1970). Firms face similar information problems. In comparison, significant amounts of resources are devoted to understanding difficult collective choices. Think of the setting up of the Intergovernmental Panel for Climate Change (IPCC) to guide decisions related to climate change.

A more robust theory emerges when positive transaction costs are introduced. When transaction costs are positive individuals cannot directly participate in policy choices and must rely on agents (politicians) whom they cannot control. Coalitions also face different levels of transaction costs of acting collectively and, therefore, will have varying capability to influence collective choices. Moreover,

far more complex institutional arrangements than simple unanimity or majority decision rules are needed to facilitate collective choices when transaction
costs are positive. These institutional arrangements partly determine the severity
of principal–agent problems and the distribution of transaction costs between
coalitions.

The positive transaction cost model highlights that coalitions influence policy
choices to a different degree because of differences in their resources and the
transaction costs they face, and because of institutional arrangements that regulate decision-making in collective choices. Therefore, policies that do not maximize social welfare may well reflect the interests of politicians, administrators, or
others well-placed to influence policy choices. Undoubtedly, these explanations
are often valid: politicians do have opportunities to trade votes, administrators
have their own interests in certain policy choices, and owners of businesses
often turn regulation to their advantage. The positive transaction cost model has
also another explanation for apparently nonmaximizing policy choices. What
appear as superior solutions in a framework that excludes transaction costs may
be inferior when transaction costs are acknowledged (Buchanan and Stubblebine,
1962; Dahlman, 1979). Indeed, all policy alternatives must be reassessed when
transaction costs are acknowledged. However, the acknowledgement of positive
transaction costs also means that uniquely optimal solutions do not exist and
therefore policy choices ultimately dissolve into distributive choices (Coase, 1960;
Calabresi, 1991; Samuels, 1992).

However, it defies logic and common sense that all deviations from standard
prescriptions from economics can be explained as the successful pursuit of
welfare-centered strategies by well-placed and powerful coalitions. After all, many
environmental policies confer diffuse benefits to large groups who are, according
to the conventional wisdom, in a weak position to advance their interests (Olson,
1971). While "diffuse" environmental interests do have their agents in environmental movements, it has been pointed out that significant environmental
benefits were created for large groups by changes in environmental policy well
before environmental organizations became influential (Elliott, Ackerman, and
Millan, 1985).

The pursuit of other-regarding and nonwelfarist goals is another possible explanation for nonmaximizing policy choices. Economists are usually resistant to this
explanation, because they are eager to extend the market model with its assumed
self- and welfare-centered agents to all realms of life. In reality, we all act on
both other-regarding and nonwelfarist values, as well as on self- and welfare-
centered ones. The character and context of choices will influence but not *determine* which values we act upon. Our choices in the market are usually informed
by self- and welfare-centered values, because markets reward it. This is the
reason for establishing markets in the first place. The pursuit of self-interest in
markets delivers collective benefits because it rewards productivity and innovation,
and because it makes collusion to fix prices and to share markets more difficult.
But markets also allow us to act upon other-regarding or nonwelfarist values
– say, when joining a consumer boycott or engaging in Green consumerism
(Paavola, 2001). Nonmarket behavior – behavior related to the family and friends

– is in turn often (though not always) informed by values other than self-centered welfarism, despite Becker's (1981) views to the contrary. The institutional context within which collective choices are made enables agents to pursue other-regarding and nonwelfarist goals to a greater extent than markets. There are good reasons for this. While the pursuit of self-interest in markets may be beneficial from the collective standpoint, this behavior does not merit the same praise where collective choices are made, as public-choice scholars have demonstrated. Yet, just like markets, nonmarket institutional settings do not determine what values can inform choices and thus they allow agents to seek their personal welfare.

Value pluralism thus characterizes collective choices and complicates them in both theory and practice. I propose to use the positive transaction cost model of collective choice as the starting-point for my argument, and then to inject the implications of value pluralism into it. This framework understands agents (both individuals and coalitions) as striving to realize their values through collective choices. In this pluralist view, some agents pursue welfare goals while others may seek nonwelfarist goals. The resources commanded by agents, the transaction costs they face, and the institutional rules that structure decision-making in collective choices determine whose values will be translated into public policy.

This pluralist theory of policy choice is not well suited for prediction, but it *is* well suited for explaining collective choices *ex post*. The reason is clear. When ethical judgments of individuals and coalitions about environmental policy are formally different, there is no way to make them commensurate – a common metric does not exist. Therefore, optimal policy choices do not exist. It is also impossible to compensate losers when they have nonwelfarist goals. Moreover, although it remains possible to make collective choices, there is no algorithm for arriving at collective choices that would be applicable in every choice situation. Collective choices simply involve deliberation to choose between the values that are to inform public policy. The characteristics of the choice problem, the constituency (the constellation of agents and coalitions with their varying goals) participating in the collective choice, and the institutions of collective choice all influence what values will be accorded public recognition after deliberation. There is no *ex ante* explanation for a collective choice, but a particular collective choice can be described and possibly explained *ex post*.

Although it is difficult to mobilize this pluralist theory of collective choice for making substantial policy prescriptions – because, from the pluralist viewpoint, solutions cannot be found but have to be worked out instead – it is still useful for political reform and policy design. However, the use of pluralist theory for these purposes hinges centrally on its implications for the performance of environmental policies.

Rethinking the Performance of Environmental Policies

Conventional environmental economics has a welfarist view of the performance of environmental policies that is based on a particular understanding of environmental problems. Environmental problems are usually conceptualized as externalities or as noncompensated physical effects between agents (Mishan, 1971; Papandreou, 1993). This conceptualization identifies an environmental problem as a deviation of resource allocation from the hypothetical ideal generated by universal, nonattenuated private property rights, and perfect markets in the absence of transaction costs (Dragun and O'Connor, 1993). Correspondingly, environmental policies are understood as instruments with which to correct resource allocation and to maximize social welfare.

This understanding of environmental problems as instances of nonoptimal allocation of (environmental) resources serves several functions in conventional environmental economics. First, it enables environmental economists to defend the status quo by arguing that the proposed policy goal or alternative does not promise an improvement in resource allocation (and thus social welfare). Second, this understanding enables environmental economists to promote policy alternatives altering the status quo by referring to a possibility of improving resource allocation (and social welfare). Third, this understanding provides a basis for an argument according to which a nonoptimal resource allocation should be corrected at the lowest possible cost in order to maximize social welfare. The performance of environmental policies is thus exhausted by their welfare consequences.

The basis of this view is problematic, because it starts with the presumption that there is a unique optimal allocation that would be achieved if only universal, nonattenuated private property rights and markets existed. There is no guarantee of the existence of such an allocation. The status quo – and the policy alternatives that would alter it – are all underpinned by different rights configurations. These rights configurations would engender different resource allocations. All of these rights configurations result in efficient allocations that have different distributive implications and are not comparable in Pareto terms (Bromley, 1989; Calabresi, 1991; Vatn and Bromley, 1994, 1997). Thus, an allocation of resources engendered by one rights configuration cannot serve as a standard with which to judge allocations based on other rights configurations. This means that private property rights – which are given effect in the analytic framework of the basic competitive model – cannot form the measuring rod for other ways of establishing rights, such as particular environmental regulations.

There is another problem with the notion of policy performance based on understanding environmental problems as instances of sub-optimal allocation of resources. Optimality does not have a clear relationship to substantive allocative outcomes that can be considered problematic by those who are affected by them. For example, the optimal allocation of water for power generation or waste disposal does not console those who would rather use water for recreation, or to protect the water for fish habitat. Similarly, the interests of an industry in need of process or cooling water are not served by the optimal allocation of water mainly for recreational uses or for the preservation of water habitat. The dissatisfied agents are often said to be able to bargain in order to change the status quo. However, the status quo may not be ideal from their viewpoint, even if they do not bargain. The distribution of income, wealth, and transaction costs may prevent them from changing the status quo in ways they would desire, or they may simply find it inequitable to have to protect their interests through markets.

The plurality of agents' interests in environmental resources and scarcity – the fundamental inability of environmental resources to cater for all interests simultaneously – engender these conflicts over the use of environmental resources (Schmid, 1987). Environmental conflicts force collective choices upon us: we have no escape from having to decide by action or inaction whose interests in environmental resources should be realized. When the nature of environmental problems *as conflicts* is acknowledged, it becomes clear that to focus on whether or not some physical effects between agents are priced is to miss the point. The choice of an environmental policy determines whose interests are realized in the resulting (optimal) allocation of resources. Environmental policies implement these collective choices by protecting certain agents' interests in environmental resources as rights, and by imposing duties on other agents (Bromley, 1991). Moral judgments are inevitable when one favors a particular rights configuration and its allocative and distributive consequences (Coase, 1960: 43; Schmid, ch. 9, this volume; Sen, 1989, 99). To reiterate, choices over environmental policies are collective judgments regarding whose interests *should* be realized and protected.

This understanding suggests that welfare judgments do not enjoy a privileged status when values are plural and in conflict with each other. Rather, the performance of environmental policies relates to what degree they achieve the aims of collective choice – the protection of certain interests in environmental resources. As discussed earlier, plural values characterize these interests. Therefore, if economic analysis wishes to be useful for making policy choices, it must characterize the performance of policy alternatives with criteria that indicate to what degree they would realize collective environmental goals. This does not preclude the use of welfare criteria, but it recommends other criteria as well. These criteria could relate to the perceived welfare or rights of nonhumans, to physical environmental outcomes that are considered intrinsically valuable, or to distributive consequences that would be engendered by policy alternatives. The criteria would also need to acknowledge the nonconsequentialist dimensions of policy choices – whether they facilitate public participation and how they position individuals *vis-à-vis* collective choices, for example.

We must also take a closer look at transaction costs, because they importantly influence the choice, implementation, and performance of environmental policies. Usually, transaction costs are understood as the costs of completing a market transaction – information costs, contracting costs, and enforcement costs (Coase, 1960). The establishment and enforcement of environmental policy also entails transaction costs. Gaining information, conducting negotiations, making collective decisions, encoding collective choices into institutional arrangements and rules, and enforcing these institutional arrangements are all costly efforts. The level and distribution (configuration) of transaction costs is jointly determined by the physical, social, and institutional contexts of the policy problem.

The physical attributes of environmental resources generate a transaction cost configuration that importantly influences the establishment of environmental policies to govern their use, as well as the implementation and outcomes of these policies. Resource attributes that increase the costs of making and enforc-ing collective choices include large size; indivisibility; rivalry of use; possibility for multiple use; mobility; fluctuating yield; impossibility of storage of resource units; and risks, uncertainties, and irreversibility related to the resource (Schmid, 1987; Gardner, Ostrom, and Walker, 1990; Schlager, Blomquist, and Tang, 1994; Feeny, Hanna, and McEvoy, 1996). This means that universal judgments concern-ing desirable institutional solutions are not possible: environmental problems differ from each other as policy problems, irrespective of values, and also require different kinds of policy responses.

The transaction cost configuration is also influenced by the social context within which resource use conflicts and collective choices take place. The number and type of involved agents, the nature and diversity of values that inform them, the nature and distribution of knowledge, and the types and amount of social capital all influence transaction costs (see Schmid, 1987; Libecap, 1995; Schlager and Blomquist, 1998). Several implications follow. First, conflicts involving a great number of heterogeneous agents with different values and imperfect knowledge are more difficult to resolve than conflicts among more homogen-ous and fewer agents. We see this in the difficult efforts to forge an international response to climate change. Moreover, some countries may be unable to resolve resource use problems such as deforestation because of lack of rule of law (Deacon, 1994): formal policies may entail prohibitively high transaction costs in implementation if compliance is not voluntary.

Finally, the level and distribution of transaction costs depends on the design of environmental policies as institutional arrangements that govern the use of environmental resources. Two aspects of institutional design merit special atten-tion here. The first is the way in which environmental policies as institutional ar-rangements organize the functions of environmental governance. The functions include exclusion of unauthorized users, regulation of resource use, monitoring of resource use, enforcement of the rules of resource use, conflict resolution and collective choice (Ostrom, 1990). Some ways of organizing environmental gov-ernance are costlier in terms of transaction costs and more difficult to implement than others: compare rules entitling individuals to a certain air quality that are enforceable in the courts, by administrative agencies, or both. The other aspect of institutional design relates to the formulation of specific institutional rules – such as the rules that regulate resource use – which may importantly influence the costs of monitoring resource use and detecting violations.

The acceptance of value pluralism and positive transaction costs do not deny the usefulness of economic analysis for environmental policy, but they suggest a somewhat different role for economics than has been customary. The accept-ance of pluralism and positive transaction costs make it impossible to present an axiomatic way to identify desirable social choices as well as to present *universal* normative judgments on the choice of policy goals and instruments. Individual and collective environmental goals are likely to vary from one situation to another, because environmental resources have different uses and the parties affected by these uses also vary. Moreover, different environmental problems structure interest conflicts differently, create different transaction cost configurations, and alter the relative influence of different interest groups on policy choices. Furthermore, policy instruments will engender different outcomes in different resource use situations and some policy instruments may realize certain goals better than others.

Judgments about the attainability of policy goals and the relative usefulness of different policy instruments must thus remain contextual when value pluralism and positive transaction costs are admitted. This means that economic analysis should be capable of particularization in addition to generalization: it should be able to improve our understanding of how particular policy goals can best be forwarded in a specific institutional, social, and physical context. Explicit and reflective incorporation of both pluralism and positive transaction costs is likely to be important for this task. This is still largely uncharted terrain for environmental economics, although the research on the governance of natural resources under common property and other institutions (see Bromley, 1999; Ostrom, 1990) has taken important steps that are likely to be applicable more broadly to environmental issues. The suggested redefinition of economic research program on the environment is not negative. First, a host of new questions are opened up for theoretical and empirical research. Second, the policy relevance of economics does not actually suffer. The acknowledgment of pluralism and contextuality encourages careful analysis, and should improve policy choices if the goodness of policy choices is judged on the basis of how they realize chosen policy goals.

Conclusions

Conventional environmental economics has focused on the welfare implications of environmental policy, ignoring its nonwelfare implications. This is unsatisfactory, because citizens care about the environment beyond its implications for their personal welfare. The omission of nonwelfarist concerns is one important reason why noneconomists are often critical of the conventional economic approach to environmental issues. I have here examined ways to accommodate a broader range of motivations in economic analysis and explored the consequences of doing so. I have suggested that the notion of rationality as strictly maximizing behavior can be relaxed so as to include all action that is consistent with the agent's reasons. This broader notion of rationality can accommodate the seeking of other-regarding and nonwelfarist goals. In essence, individuals can be understood to seek to realize their values. This means that some agents may be concerned about their own welfare, others about the goals they consider intrinsically valuable, and still others about right or virtuous ways to act. When conflicting values are plural and incommensurable, there are no socially optimal choices. Collective choices can still be explained *ex post*, but their prediction may not be possible.

The pluralist theory has a number of important implications for the analysis and making of collective decisions, especially when coupled with the recognition of imperfect information and positive transaction costs. The pluralist theory emphasizes the role of collective decision-making as a process of learning and preference formation, in which participation and facilitating procedures are important. It also suggests that we might group collective choices into classes associated with particular decision-making procedures. Decisions that are clouded by imperfect information and characterized by conflicting ethical standpoints – the use of genetically modified organisms, for example – are different from decisions that can be made with adequate information and be based on consensual values. Choices that are mainly characterized by imperfect and uncertain information are also different from choices mainly characterized by conflict of values. We sometimes need to act on uncertain information and adopt measures such as the precautionary principle to assist us in making decisions. Ethical dilemmas also need to be resolved, but new information and learning will not do it. An ability to participate in the policy debates and choices and explicit justification of policy choices in moral terms may do it. This does not mean denial of welfare considerations in policy choices. It means, instead, that welfare considerations must be justifiable and *justified* as guiding ethical principles in the context of particular policy choices. If welfare concerns are not justifiable, other ethical grounds for choice are needed.

The pluralist theory also has implications for understanding and analyzing the performance of environmental policies. It indicates that we should focus on how policy alternatives realize collectively agreed goals, instead of focusing on their welfare consequences. For example, this can translate into the use of multiple criteria for characterizing policy performance – criteria that have to track real consequences such as changes in the state of the environment that are of importance for agents. It also means attending to the procedural and rights dimensions of policy alternatives. The pluralist theory also calls for placing more emphasis on the implementation of environmental policies in the real world, instead of focusing on theoretical implications in an ideal world. One step in this direction would be to acknowledge and analyze the effects of positive transaction costs on the attainment of policy goals. At the level of policy-making, the theory calls for recognizing that policy problems are different both in degree and in kind. Careful analysis and crafting of institutional solutions is needed in this kind of a context. This has, perhaps, been lost more in the economic debates on policy instruments than in the actual process of making public policy.

References

Ackerman, B. A. and Hassler, W. T. 1981: *Clean Coal/Dirty Air*. New Haven, CN: Yale University Press.

Akerlof, G. 1970: "The market for lemons": quality uncertainty and the market mechanism. *Quarterly Journal of Economics*, 84, 488–500.

Anderson, E. 1993: *Value in Ethics and Economics*. Cambridge, MA: Harvard University Press.

Axelrod, R. 1984: *The Evolution of Cooperation*. New York: Basic Books.

Becker, G. S. 1981: *A Treatise on the Family*. Cambridge, MA: Harvard University Press.

Bromley, D. W. 1989: *Economic Interests and Institutions: The Conceptual Foundations of Public Policy*. Oxford: Blackwell.

Bromley, D. W. 1991: *Environment and Economy: Property Rights and Public Policy*. Cambridge, MA: Blackwell.

Bromley, D. W. 1999: *Sustaining Development: Environmental Resources in Developing Countries*. Cheltenham, UK: Edward Elgar.

Buchanan, J. M. and Stubblebine, W. C. 1962: Externality. *Economica*, 29, 371–84.

Calabresi, G. 1991: The pointlessness of Pareto: carrying Coase further. *Yale Law Journal*, 100, 1211–37.

Coase, R. H. 1960: The problem of social cost. *Journal of Law and Economics*, 3, 1–44.

Cooter, R. and Rappoport, P. 1984: Were the ordinalists wrong about welfare economics? *Journal of Economic Literature*, 22, 507–30.

Dahlman, C. J. 1979: The problem of externality. *Journal of Law and Economics*, 22, 141–62.

Deacon, R. T. 1994: Deforestation and the rule of law in a cross-section of countries. *Land Economics*, 70, 414–30.

Dragun, A. K. and O'Connor, M. P. 1993: Property rights, public choice, and Pigouvianism. *Journal of Post Keynesian Economics*, 16, 127–52.

Eggertsson, T. 1990: *Economic Behavior and Institutions*. Cambridge: Cambridge University Press.

Elliott, E. D., Ackerman, B. A., and Millan, J. C. 1985: Toward a theory of statutory evolution: the federalization of environmental law. *Journal of Law, Economics, and Organization*, 1, 313–40.

Elster, J. 1983: *Sour Grapes: Studies in the Subversion of Rationality*. Cambridge: Cambridge University Press.

Feeny, D. Hanna, S., and McEvoy, A. F. 1996: Questioning the assumptions of the "Tragedy of the Commons" model of fisheries. *Land Economics*, 72, 187–205.

Freeman III, A. M. 1990: Water pollution policy. In P. R. Portney (ed.), *Public Policies for Environmental Protection*. Washington, DC: Resources for the Future, 97–149.

Gardner, R. Ostrom, E., and Walker, J. M. 1990: The nature of common-pool resource problems. *Rationality and Society*, 2, 335–58.

Georgescu-Roegen, N. 1968: Utility. In D. L. Sills (ed.), *International Encyclopedia of the Social Sciences*, Vol. 16. New York: Macmillan/The Free Press, 236–67.

Hahn, R. W. 1989: Economic prescriptions for environmental problems: how the patient followed the doctor's orders. *Journal of Economic Perspectives*, 3, 95–114.

Hicks, J. and Allen, R. G. D. 1934: A reconsideration of the theory of value. *Economica*, 1, 52–76, 196–219.

Huber, P. 1983: The old–new division in risk regulation. *Virginia Law Review*, 69, 1025–107.

Jorgensen, B. S., Syme, G. J., Bishop, B. J., and Nancarrow, B. E. 1999: Protest responses in contingent valuation. *Environmental and Resource Economics*, 14, 131–50.

Kavka, G. S. 1991: Is individual choice less problematic than collective choice? *Economics and Philosophy*, 7, 143–65.

Libecap, G. D. 1995: The conditions for successful collective action. In R. O. Keohane and E. Ostrom (eds.), *Local Commons and Global Interdependence: Heterogeneity and Cooperation in Two Domains*. London: Sage, 161–90.

Mishan, E. J. 1971: The postwar literature on externalities: an interpretative article. *Journal of Economic Literature*, 9, 1–28.

Olson, M. 1971: *The Logic of Collective Action: Public Goods and the Theory of Groups*, 2nd edn. Cambridge, MA: Harvard University Press.

Ostrom, E. 1990: *Governing the Commons: The Evolution of Institutions for Collective Action.* Cambridge: Cambridge University Press.

Paavola, J. 2001: Towards sustainable consumption: economics and ethical concerns for the environment in consumer choices. *Review of Social Economy,* 59, 227–48.

Palmer, K., Oates, W. E., and Portney, P. R. 1995: Tightening environmental standards: the benefit–cost or the no-cost paradigm? *Journal of Economic Perspectives,* 9, 119–32.

Papandreou, A. 1993: *Externality and Institutions.* Oxford: Oxford University Press.

Rawls, J. A. 1971: *A Theory of Justice.* Cambridge, MA: Harvard University Press.

Samuels, W. J. 1992: The pervasive proposition, "What is, is and ought to be": a critique. In W. S. Millberg (ed.), *The Megacorp and Macrodynamics: Essays in Memory of Alfred Eichner.* Armonk, NY: M. E. Sharpe, 273–85.

Schlager, E. and Blomquist W. 1998: Heterogeneity and common pool resource management. In E. Tusak Loehman and D. M. Kilgour (eds.), *Designing Institutions for Environmental and Resource Management.* Cheltenham, UK: Edward Elgar, 101–12.

Schlager, E., Blomquist, W., and Tang, S. Y. 1994: Mobile flows, storage, and self-organized institutions for governing common-pool resources. *Land Economics,* 70, 294–317.

Schmid, A. A. 1987: *Property, Power, and Public Choice: An Inquiry into Law and Economics,* 2nd edn. New York: Praeger.

Sen, A. K. 1973: Behavior and the concept of preference. *Economica,* 40, 241–59.

Sen, A. K. 1989: The moral standing of the market. In D. Helm (ed.), *The Economic Borders of the State.* Oxford: Oxford University Press, 92–109.

Sen, A. K. 1991: Utility: ideas and terminology. *Economics and Philosophy,* 7, 277–83.

Simon, H. A. 1978: Rationality as a process and product of thought. *American Economic Review,* 68, 1–16.

Spash, C. L. and Hanley, N. 1995: Preferences, information, and biodiversity preservation. *Ecological Economics,* 12, 191–208.

Vatn, A. and Bromley, D. W. 1994: Choices without prices without apologies. *Journal of Environmental Economics and Management,* 262, 129–48.

Vatn, A. and Bromley, D. W. 1997: Externalities – a market model failure. *Environmental and Resource Economics,* 9, 135–51.

7

What Should We Do with Inconsistent, Nonwelfaristic, and Undeveloped Preferences?

Olof Johansson-Stenman[1]

Policy conclusions in applied welfare economics, such as environmental economics, normally presuppose that people have well-defined and consistent preferences. However, there is empirical evidence that people's preferences with respect to environmental goods are often far from complete and highly context-dependent. This chapter will discuss what are the policy implications when people's preferences are not consistent or even developed. Furthermore, it is usually assumed that what matters for public policy is people's utilities, and nothing else. However, this assumption can be questioned in the environmental field, and there is evidence that some people hold nonanthropocentric ethical views. We will therefore examine the possibility that the environment, or animal welfare, may matter intrinsically from a social point of view, and not only instrumentally through people's utility functions.

Empirical tests have often rejected standard economic assumptions with respect to people's preferences and behavior. Although this has been discussed intensively in recent years (Thaler, 1992, 2000; Conlisk, 1996), the practical consequences for applied economics, such as benefit–cost analysis, appear minor so far. Why is that? There exist at least four ways to defend the prevailing practice. First, applied work becomes too complicated to undertake if the standard assumptions are not maintained. Second, policy recommendations become less straightforward to derive. Third, the standard assumptions – although, strictly speaking, incorrect – provide an *approximate* picture that is sufficiently accurate for the tasks at stake. Fourth, we must maintain our most fundamental assumptions, because otherwise there is nothing left of economics. The fourth reason is, of

[1] I am grateful for constructive comments from Jouni Paavola and Daniel Bromley, who provided very detailed and insightful comments. I have also received useful comments from Fredrik Carlsson, and financial support from the Swedish Transport and Communications Research Board (KFB, or Vinnova) is gratefully acknowledged. The usual disclaimer applies.

course, against what is (or should be) the most fundamental characteristics of science, critical thinking and work to replace existing theories with better ones, and can therefore be dismissed as unacceptable.[2] The previous three arguments can, however, be taken seriously. It is true that anomalies and deviations from the predictions of the standard theory are sometimes small. There is also a tradeoff between simplicity and relevance in economics (as in all scholarship), and without some simplifying assumptions we could not draw *any* conclusions. So, in some cases the standard theory appears to be the appropriate one to use. But sometimes it is not. There is evidence that expressed preferences and observed behavior with respect to the environment and risky choices deviate widely from what the standard theory predicts. Therefore, it is also important to consider cases where people's preferences are *not* perfectly informed, consistent, and fully developed with regard to goods, such as environmental goods. For example, how should we handle situations in which people's risk-perceptions are biased? Should public policy be based on people's preferences or expected welfare consequences?

Mainstream normative economics is usually based on either a specific ethical theory, such as utilitarianism, or on some weaker minimum requirements of the social objective. In particular, it is often assumed that the Pareto principle must not be violated in terms of individual utilities. This implies that if the social objective is expressed as a social welfare function (SWF) to be maximized, this SWF must solely depend on individual utilities. Using the terminology of Sen (1979), such a SWF is *welfaristic*. Environmental quality or animal welfare may affect social welfare, but only indirectly through the individual utility functions. This chapter discusses how a benevolent policy-maker could and should act based on a fundamental, possibly *non*welfaristic, ethical principle, meaning that we allow for more general SWFs – where, for example, animal welfare and environmental quality matter *per se*, irrespective of whether or not people obtain any utility from them.

In what follows, the first section discusses individual welfare, as a measure of individual well-being, and preferences, as reflected in revealed choices or stated preferences, as "ends" for the government to pursue. It is concluded that welfare, rather than preferences, is an appropriate end for a reasonable consequentialist ethical theory. The second section discusses whether individual welfare should be the only end, or whether there might be other possible ends for the government. On the basis of empirical evidence, it is argued that many people seem to value the environment intrinsically, and that these views should also be reflected in social decisions. The third section discusses what to do when preferences are inconsistent or irrational because of cognitive limitations, biased risk perceptions, cognitive dissonance, or myopic behavior. The fourth section discusses if – and if so, how – insights on preference formation are important from a welfare standpoint. It is argued that the actual preferences that we are looking for are those that reflect welfare as closely as possible, and in addition provide the most accurate information about the agents' view of what should

---

[2] Even though such provincial thinking is not very often found in print, it is often put forward, also by respectable economists, in day-to-day talk.

intrinsically matter. Hence, environmental valuation methods should explicitly be constructed to elicit such preferences. The final section discusses how the insights from this chapter might be used in policy-making.

Utility, Preferences, and Well-Being

In what is often denoted "positive" economics, such as consumer demand analysis, we are interested in explaining and describing observed phenomena – for example, consumption patterns – in terms of price and income elasticities. The (assumed unobservable) utility function is typically defined implicitly as a function (with a small amount of imposed structure) that is maximized by the observed consumption pattern, or more generally by individuals' actions, following Samuelson (1938). There is an infinite number of utility functions that can explain a consumption pattern. For example, if a well-behaved[3] utility function $u = U(x_1, x_2, \ldots, x_n)$ explains a certain consumption pattern, it is straightforward to show that any monotonically increasing transformation $f(u)$ is consistent with the same pattern (see Samuelson, 1947; Graaf, 1957).[4] The utilities, representing preferences, can thus not be interpreted in a cardinal way, and statements such as "utility is concave in income" are meaningless.

On the other hand, in what is often described as "normative" economics, including policy analysis, utility (cardinal or ordinal) is used to represent individual welfare or well-being. In some cases there is no need to make a distinction between these two different uses of the term "utility." This would be the case when people's choices are explained *solely* by the maximization of individual welfare. However, sometimes this is not the case. As remarked by Broome, "Welfare economists move, almost without noticing it, between saying a person prefers one thing to another and saying she is better off with the first than with the second" (1999: 4). Here we will distinguish between utility as representing preferences, on the one hand, and welfare on the other. The preferences are defined by choices, actual or what they would have been in a real choice situation (Broome, 1999), whereas welfare is used interchangeably with well-being, a notion that may be interpreted more broadly than individual happiness, however.

Many philosophers, psychologists, and also some economists have criticized the narrow view of individuals as concerned exclusively with utility maximization. However, many (most?) economists consider the often quoted statement by Gary Becker in his Nobel Lecture on "the economic way" to provide an effective end to this discussion: "Individuals maximize utility *as they perceive it*, whether they be selfish, altruistic, loyal, spiteful, or masochistic" (Becker, 1993: 386, italics in original). For example, altruistic concerns are consistent with, and may be modeled within the framework of, utility maximization. And if we think of utility solely as something that is implicitly maximized in order to be consistent

---

[3] That is, increasing, strictly quasi-concave, and twice continuously differentiable in the arguments.
[4] For example, where $f(u) = \exp[U(x_1, x_2, \ldots, x_n)]$ or $f(u) = \ln[U(x_1, x_2, \ldots, x_n)]$.

with actual behavior, then the statement is of course tautologically true, but also virtually meaningless. As expressed by Samuelson more than 50 years ago:

> Thus, the consumer's market behavior is explained in terms of preferences, which are in turn defined only by behavior. The result can very easily be circular, and in many formulations undoubtedly is. Often nothing more is stated than the conclusion that people behave as they behave, a theorem which has no empirical implication, since it contains no hypothesis and is consistent with all conceivable behavior, while refutable by none. (Samuelson, 1947: 91)

Therefore, to say that people are utility-maximizers is like saying that football players are *Sex*-minimizers, where *Sex* is defined implicitly as something that if it is minimized is consistent with the actual behavior on the field. The only information we could get from this statement is that the football players' behavior is not completely random, something that is about as obvious (at least for someone who knows the rules) for football players as it is for consumers. If, on the other hand, utility is seen as a measure of individual well-being, then Becker's statement is certainly not trivially true, but a meaningful hypothesis which in principle is testable and refutable.

The utility function used in policy-oriented economic analysis, such as environmental valuation, is often (implicitly) assumed to be both a measure of well-being and of choice simultaneously. This is often not recognized, which can partly explain the often confused discussions on how to interpret environmental-valuation results (see Broome, 1991a, 1999; Sen, 1991; Johansson-Stenman, 1998). As repeatedly emphasized by Sen (1977, 1985, 1987), whether or not people maximize their own well-being is ultimately an empirical question, although a difficult one. Thus, if well-being is *defined* independently of choice, and an individual chooses A instead of B, it certainly does not follow that the individual's well-being must be larger with A than with B. First, people simply make mistakes. Second, people may prefer to sacrifice some of their own well-being in order to attain another end, such as their childrens' well-being. The fact that an individual may derive utility from being kind to another person (sympathy, in Sen's terminology) does not imply that all kindness is caused by a utility-maximizing behavior. The behavior may in part be due to some other motive, such as the welfare of another person. Hence, the statement by Becker, although perhaps superficially appealing, is not at all obvious, unless utility is defined in a way so as to make the statement tautologically true.[5]

In the special case when preferences are a perfect measure of welfare, it does of course not matter whether we choose preferences or welfare as our unit of analysis. But when preferences and welfare differ, should public decision-making be intrinsically concerned with preferences or welfare? If we limit our analysis to consequentialism, and rule out right-based ethics, it appears straightforward that welfare, rather than preferences, is good in itself. Things may be chosen because

---

[5] In defense of Becker, one should note that he himself does not seem to view individual utility-maximization as a *hypothesis* but, rather, as a *method* of analysis.

they are good, but they are not good because they are chosen.[6] This position is very similar to the betterness principle argued for by Broome (1991b, 1999), and also related (although not identical) to the view of Harsanyi (1982, 1995, 1997), who has repeatedly argued that what should matter in social decision-making is the *true* or *informed* preferences; that is, the preferences that a rational individual equipped with perfect information would have, rather than actual preferences based on partial information.[7] Still, as will be argued, not even perfect information and the use of an infinite cognitive capacity will guarantee that a person's preferences will reflect his or her welfare in cases in which a person deliberately chooses to maximize some other end than his or her own welfare.

It is also difficult to consistently argue in favor of preferences in cases where preferences and welfare differ. For example, how should one argue in the common situations in which individuals prefer to have restricted choices (Akerlof, 1991), or that the government should choose for them? And why would they prefer to have restricted choices in the first place, if not because what is intrinsically important is welfare, and they believe that welfare will be higher with restricted choices (including possible cognitive effort and so on)?

Beyond Anthropocentric Welfarism?

The standard economic model does not distinguish between welfare and preferences, and uses utility to represent both. It also implicitly assumes that there is nothing intrinsically important besides individual (human) utilities. Here, on the contrary, evidence will be presented that many people seem to value the environment intrinsically, and it is argued that the government should take such views seriously.

An anthropocentric and welfaristic social welfare function can be written as $W = w(u^1, u^2, \ldots, u^n)$, where W is social welfare and u^i is utility for individual i. Environmental quality or animal welfare may influence social welfare, but only indirectly through the individual utility functions as follows:

$$W = w(u^1(\mathbf{x}^1, \mathbf{z}) u^2(\mathbf{x}^2, \mathbf{z}), \ldots, u^n(\mathbf{x}^n, \mathbf{z})), \tag{1}$$

where $\mathbf{x}^j$ is j's consumption of a vector of private goods, and where $\mathbf{z}$ is a vector of public goods, including various aspects of environmental quality or animal welfare. For example, the suffering of a particular animal species may affect social welfare through altruistic (or sympathetic) concern in one or many individuals'

[6] This is, of course, not to say that preferences should not be considered in public decision-making, since preferences may often be much easier to observe, and they may in many cases be closely correlated with welfare.

[7] Ng (1999) goes one step further by arguing that the logical consequence of Harsanyi's arguments is that the government should be concerned solely with individual happiness. This view is not necessary for the arguments here, although most people would probably agree that happiness is an important element of welfare, or well-being.

utility functions. Although such an anthropocentric view dominates in welfare economics, it is not very often clearly expressed in plain English. Baxter (1974) is an exception:

> Penguins are important because people enjoy seeing them walk about rocks; and furthermore, the well-being of people would be less impaired by halting use of DDT than by giving up penguins. In short, my observations about environmental problems will be people-oriented, as are my criteria. I have no interest in preserving penguins for their own sake. (Baxter, 1974: 5)

Although it is clear that Baxter holds purely anthropocentric values and that many other economists probably also do so, empirical findings suggest that many other people do not. Stevens et al. (1991), Spash and Hanley (1995), Russell et al. (1999), and others have found that many people seem not only to value human well-being, but also nature in itself, including animal welfare. A more general SWF reflecting such values can be written as follows:

$$W = w(u^1(\mathbf{x}^1, \mathbf{z})u^2(\mathbf{x}^2, \mathbf{z}), \ldots, u^n(\mathbf{x}^n, \mathbf{z}), \mathbf{z}). \tag{2}$$

The difference between (1) and (2) is that $\mathbf{z}$ is an argument by itself in the SWF in (2), irrespective of individual utilities. In the latter case we allow for the possibility that social welfare may decrease due to animal suffering even if no human being knows (or cares) about it. The empirical results by Russel et al. are particularly interesting, since one purpose in that study was to trigger private, social, and ecological preferences by changing the framing of questions associated with the valuation of a recreation area. In follow-up questions respondents were asked about motives for their choices, including questions about whether the well-being of other people or the value of nature *per se* had affected their responses. Most respondents gave "much weight" to the consequences for themselves and for their family. Less than half of the respondents gave "much weight" to the effects for other visitors. The consequences for "the flora and fauna in the forest," irrespective of their value for the people, were given "much weight" by more than four out of five respondents in every framing, including the private one. This should be contrasted with the mainstream assumption, which predicts that nobody would be influenced by such motives.

Given that many people think that the environment and animal welfare should be given *some* weight in public decision-making, should the government respect these views? More generally, should public policy be based exclusively on people's opinions? The answer is not self-evident. For example, Pigou (1929) argued that it is the duty of the government to protect the interest of future generations from the current generation's selfish shortsightedness. Marglin (1963), on the other hand, argues that the government should only acknowledge the preferences of present individuals (who may, of course, have preferences for people in the future). The question may be difficult to answer conclusively, since it relates to conflicts between fundamental *values* and fundamental *principles* for democratic decision-making. Yet it is less problematic in our specific case. If most people prefer the government to *also* consider the environment or animal welfare *per se*, in addition to their own welfare, it seems reasonable for the government to do so. It would seem strange if the government would not put *any* weight on the value of nature, or on avoiding animal suffering, in tradeoffs between human welfare versus nature and animal welfare, if that is what most people want it to do.

It should be emphasized that in order to accept that not only human well-being has an intrinsic value one need not go as far as some utilitarians such as Singer (1975, 1979), who argue that all suffering should count equally (per suffering unit). Nor does one have to accept that environmental entities have absolute rights (see, for example, O'Neill, 1997). It is sufficient that in tradeoffs between human well-being on the one hand, and nature or animal well-being on the other, the weights given to the latter are larger than zero.

Imperfect Information and Inconsistent or Irrational Preferences

Due to limited or misleading information, people often make decisions that they will later regret. The appropriate policy response is not obvious, however. One could argue that the appropriate response is to provide more, and more easily accessible, information for individuals to enable them to make informed and rational decisions. However, although information provision is an appropriate task for the government, it has its limitations. First, it is well known that many people, for various reasons, simply do not trust publicly provided information (Slovic, 2000). Second, even if they do, it would often be extremely expensive to provide all citizens with perfect information. This suggests that there is a tradeoff between welfare costs due to imperfect information, on the one hand, and costs associated with the provision of better information on the other. Third, even if perfect information could be provided, people have limited cognitive capacity and time to process the information (see Conlisk, 1996). Thus, even though there is a role for public information provision, one must still decide how to deal with situations in which people have imperfect information or limited cognitive capacity. For example, it is well-known that people have difficulties in dealing with stochastic problems: they often overestimate small probabilities and underestimate large ones (Slovic, 2000; Viscusi, 1992, 1998). This is important, because environmental policy is largely a problem of how to deal with risk; for example, since our knowledge of the ecosystem effects of environmental pollution is very limited.

Pollak (1998) recently concluded a paper by saying that it is still an open issue how policy-makers should react to information about systematic biases in individual risk perception. He argues that, "Utilitarians – and most welfare economists and policy analysts approach public policy from a utilitarian perspective – should consider whose beliefs (the public's or the experts') should be used to calculate expected benefits" (Pollak, 1998). Here it is argued that preferences are important mainly because they provide information about welfare (or other ethical goals), and that we are ultimately interested in welfare. Practical policy conclusions may be less straightforward, however. Assume unrealistically[8] that we know individuals' preferences and their cardinal welfare functions perfectly, and that individuals overestimate a certain risk. It would still remain ambiguous whether an efficient (in terms of welfare) risk-reducing policy measure should be under- or overprovided relative to the first-best (with full information) efficiency rule, in terms of preferences (Johansson-Stenman, 2000). This is because the *marginal* change in subjective risk as a function of objective risk may be smaller than one, even when the subjective risk is larger than the objective risk. This would be the result if the subjective risk is higher than the objective one at low-risk levels, and lower at high-risk levels. This seems to be the standard case in the literature (Viscusi, 1992: 139–40). Thus, even if people overestimate a risk, it does not follow that their WTP for risk *reduction* would be larger compared to the case with perfect information (and vice versa).

Besides cognitive limitations, there are other reasons, such as limited information and cognitive dissonance, why individual preferences are sometimes poor indicators of welfare. Consider a neighborhood with enhanced radon levels in drinking water, and where people are confronted with an increased risk of contracting lung cancer. Assume that individuals are first not aware of a link between radon concentrations and the probability of getting lung cancer, and that this information is provided at a specific moment in time. Before this moment, the individuals' WTPs for measures to decrease the radon level are approximately zero. After providing the information, the WTPs would increase. Most people would probably agree that the appropriate response should not be based on the preferences (reflected in their zero or very low WTPs) that prevailed before the disclosure of information. This is consistent with the arguments made so far, since the (expected) welfare increase of the response is arguably the same before and after the information is provided, even though the expressed preferences are not. We clearly cannot use the uninformed preferences as a basis for public decision-making. But can we use the after-information preferences? Possibly, but not even this is obvious.

Consider the influence of cognitive dissonance (Festinger, 1957; Akerlof and Dickens, 1982). According to this well-established psychological theory, individuals try to avoid or decrease the "dissonance" between the real circumstances and their view of them. Often, this is perfectly compatible with standard theory of rational choice, so that when a person dislikes a certain event, he or she tries to change it so that he or she would like it. For example, if you dislike your car, and if you are not poor, you will replace it with a car that you like. Sometimes, however, it is difficult, if not impossible, to change the real circumstances – such as radon in drinking water. Assume that people have lived in their houses and neighborhoods for decades and cannot move because of prohibitive costs. Either they can fully incorporate the new information and accept that their cumulative water consumption will increase their risk of getting lung cancer, or they may

---

[8] Survey-based methods are particularly problematic and are connected with many serious problems in the area of risk-valuation. See, for example, Beattie et al. (1998), Carthy et al. (1998), and Jones-Lee and Loomes (1997).

modify the conclusions toward a view that their added risk is negligible, or at least not that large. Since they cannot change the real circumstances of the past, changing their views may be tempting. The WTPs for measures to decrease radon levels would then also decrease. But the expected welfare loss due to the increased risk of getting cancer has of course not changed. Therefore, not even the fully informed preferences are always useful for public decision-making, since they may not reflect welfare accurately. The radon example[9] is not as hypothetical as it might seem, and there is empirical evidence that people underestimate the risk associated with enhanced radon levels (see Pollak, 1998; Slovic, 2000: ch. 16). Cognitive dissonance may be one explanation.

We have concluded so far that people's subjective risk perceptions and economic valuations may be biased for various reasons. But risks are often difficult to estimate for experts as well, and there is also a possibility of scientific bias. Consider again the theory of cognitive dissonance, but this time applied to scientists. Assume that you are an expert on global computable general equilibrium models of climate change, that you have invested a lot of time and effort to obtain this knowledge, and that you have much detailed data on natural scientists' best-guess scenarios. You can use the data, derive and compute your results, and claim that your results are important for policy. If you at all reflect about uncertainty, you will conclude that the inclusion of such considerations does not change anything essential. Alternatively, you can conclude that it is doubtful whether any useful policy conclusions could be drawn from the results, since important low-probability catastrophic events are not included in the model. The first alternative is presumably more attractive to the researcher, implying an incentive for the researcher to modify his or her view of reality toward this direction.

The difficulty of dealing with low-probability catastrophic events in a systematic way is by no means limited to economists. Therefore, the reason for the deviation between the experts' and lay-persons' risk perceptions may not solely be that the uninformed public exaggerates the possible low-probability catastrophic events. Experts may correspondingly underestimate the expected consequences of such outcomes because they have no good tools with which to quantify them. It seems that, within each discipline, we tend to believe that the tools we have at our disposal are in general suitable for the problems we analyze.

Hence, cognitive dissonance and the creation of a positive self-image may imply that we consider methods and theories that we do not understand to be less important. Clearly, no one can understand everything. But it is more pleasant to think that what we know is the really important part, rather than to think that the most important part is known by others. Most researchers who have tried to engage in interdisciplinary work can presumably recognize these and similar problems.

---

[9] Alternatively, assume that you live in an area that is vulnerable to earthquakes and that, for some reason, you cannot move. The real choice is not about changing these circumstances, but about whether you should change your view of these circumstances. Either you will live in an area which you know is very vulnerable to earthquakes, or you will live in the same area but begin to doubt that it really is all that vulnerable.

Another problem is the tendency to discount the future to an irrational extent, a phenomenon that remains a nonconventional assumption in economic theory.[10] Still, it has been discussed by prominent economists such as Harrod, Pigou, and Ramsey, who colorfully denoted the phenomenon "the conquest of reason by passion" (Harrod, 1948: 40), "the faulty telescopic faculty" (Pigou, 1929: 25), and the "weakness of imagination" (Ramsey, 1928: 543). If we, again, are interested in welfare rather than preferences, such myopic behavior could be an argument in favor of compulsory pension savings and health insurance, for example. The possibility of shortsightedness is also a reality in the environmental field, and not only related to individual decisions. Indeed, the shortsightedness of firms and politicians is much recognized in the discussions on public choice.

The scientific community may also suffer from "faulty telescopic faculty." For example, the problem of rapidly increasing antibiotic resistance has not been dealt with adequately, although research on it has increased recently. The strong link in medical research to commercial interests may be one reason. The "external" costs associated with an excessive antibiotic use are not, or at least very poorly, "internalized." Still, this is hardly the only factor, since the social sciences – with much looser links to commercial interests – seem to have shown even less interest and ability to deal with antibiotic resistance.[11] Thus, rather than being primarily an effect of commercial interests, it may be that academic research is poorly adapted to systematically deal with future problems with a stochastic nature. The economics of science (see Stephan, 1996) suggests several mechanisms of influence that work through the academic reward system (including status and esteem), and through the funding system. Stephan argues that "the grant system [. . .] encourages scientists to choose sure(r) bet short term projects that in the longer run may have lower social value" (1996: 1226). It may be difficult to change these mechanisms directly, but an increased awareness of their problems among policy-makers and academics might mitigate their consequences. Thus, policy-makers must be aware of the fact that agents, including academics, may sometimes suffer from various degrees of shortsightedness.

Undeveloped Preferences and Preference Formation

Most people have limited experience in assigning monetary values to environmental goods, because this requirement does not usually occur in everyday life. One may also question whether people have preferences for all environmental goods. For example, can we have preferences for a species that is threatened by extinction, if we never knew it existed? If we have preferences for the species, they must have been created rather quickly and are likely to be far from stable

---

[10] For an often-quoted paper that supports the view that consumers tend to apply an inefficiently high internal discount rate, see Hausman (1979).

[11] At the time of writing, I am aware of only two published academic papers in economics journals on antibiotic resistance (Brown and Layton, 1996; Doessel, 1998). This can be seen as an indication of an inefficient institutional and/or incentive structure within the academic knowledge-producing sector.

over time. Therefore, it is important to take the process of preference *formation* seriously. On the other hand, environmental valuation should focus on individual welfare rather than on preferences. Moreover, individual welfare associated with environmental goods may be more stable than the preferences. Yet another important task for valuation studies is to elicit people's ethical views about other ends besides human well-being.

The standard assumption in the contingent valuation literature (Mitchell and Carson, 1989) is that people know their complete preferences with respect to all goods, and that economists' role is simply to elicit them. An alternative view – more common among psychologists – is that we have developed preferences for only a few familiar goods, and that in most circumstances we use heuristic choice rules (Tversky and Kahneman, 1974; Kahneman and Tversky, 1979). This view is supported by experiments that demonstrate *preference reversals*, which happen when people prefer A over B but are still willing to pay more for B than for A (Slovic and Lichtenstein, 1983; Tversky et al., 1990). For example, Gregory et al. (1993) found that in a choice between improved computer equipment and improved air quality, most people chose improved air quality. Yet most people had a higher WTP for the computer improvement than for the air quality improvement. There is also evidence of an *endowment effect* – people demand more in compensation to give up a good than they would be willing to pay to get it – and the related phenomena *loss aversion* (Kahneman and Tversky, 1979) and *status quo bias* (Samuelson and Zeckhauser, 1988). Finally, there is evidence that putting a monetary value on an environmental change is a cognitively demanding task. Therefore, people tend to use context-dependent heuristic choice rules when confronting such a task. This makes their responses difficult to interpret (Schkade and Payne, 1994; Vatn and Bromley, 1994).

As argued by Gregory et al., the appropriate role of CV practitioners may thus be "not as archeologists, carefully uncovering what is there, but as architects, working to build a defensible expression of value" (1993: 179). Even if people have no developed preferences for environmental goods, their welfare may depend on them. Their (individual) welfare functions may also be more stable than their preferences. Moreover, even if expressed preferences would not reflect an acceptable ethical end in themselves, they are important for at least two reasons. First, preferences may provide a crude estimate of welfare. Second, they may provide useful information about other ends in addition to welfare. For example, even if we do not know about water contamination, our welfare may depend on it directly in terms of health effects or indirectly through deterministic or low-probability stochastic ecosystem effects. Hence, when preferences are context-dependent, we should be looking for those preferences that reflect welfare as closely as possible, and in addition separately provide information about people's view of what should intrinsically matter. Unfortunately, most valuation studies have not made this distinction and thus cannot inform us about the respondents' ethical values. Whether a method based on multi-attribute utility theory as proposed by Gregory et al. (1993), Gregory and Slovic (1997), and Slovic (1995) is appropriate for the task is an interesting but nontrivial issue, beyond the scope of this chapter. The method requires a lot of resources per respondent, which implies a substantially lower number of respondents than in CVM. Indeed, as expressed by Gregory et al., "Depth of value analysis is substituted for breadth of population sampling" (1993: 189). Still, the method promises results that may also improve conventional stated-preference methods, particularly with respect to cognitive difficulties and preference construction.

Thus, it is clear that preferences are endogenous and may change due to the framing and circumstances, as has been shown both in an experimental setting and by real-life experiences. However, the endogeneity and instability of preferences may not be a major problem *in principle*, since we are ultimately interested in welfare rather than preferences. Nevertheless, the endogeneity and instability of preferences may translate into large problems *in practice*, since methods for revealing or eliciting preferences are often important sources of information in order to determine or estimate welfare.

But although welfare is not as labile as preferences, also (individual) welfare functions are endogenous and may change over time for various reasons, such as habit formation or addiction. Consider a person who is severely ill, but who did not know this until recently. The new information probably changed his or her preferences, but not the expected welfare associated with the appropriate medical treatment *per se*. But the information also brings anxiety and mental suffering. Assuming that these emotions can be reduced by therapy, it is clear that the new information has also changed the welfare function. Before the new information there was no anxiety and hence no possibility of reducing it with therapy. Public decision-makers should consider ways to reduce suffering given the current state of the world, and not the possibilities of last week. This example shows that when welfare functions change over time we should, in principle, be concerned with instantaneous welfare.

In the context of environmental valuation, it is clear that many people's welfare relates to other values in addition to the so-called use values, such as their own actual or hypothetical contribution to the existence of environmental goods (Andreoni, 1990; Kahneman and Knetsch, 1992). There is no reason why these welfare effects should not count for social decision-making: moral satisfaction is as real as satisfaction that has other sources. However, the purpose of a CV study is not primarily to estimate respondents' instantaneous welfare *from responding to CV questions* but, rather, to elicit responses that are extendable also to *outside the survey context*. If the moral satisfaction primarily occurs when responding to the survey questions, those who do not belong to the sample would not obtain this improvement in welfare. This is an area where confusion still prevails. For example, in a critical comment to Kahneman and Knetsch, Harrison (1992) argued that the *motive* for the respondents' utilities is irrelevant: "I call my utility 'jolly'. What you call your utility is . . . your business" (Harrison, 1992: 150). From a benefit–cost standpoint this is incorrect: the motive behind the preferences and welfare matters a great deal (Johansson-Stenman, 1998).[12]

---

[12] Due to a somewhat unconventional editorial policy, the reply to Harrison by Kahneman and Knetsch was never accepted for publication in *JEEM*. However, the authors did send out letters to all subscribers, telling them that the reply was available personally, and many people received a copy in this way.

There is evidence that so-called nonuse values play a dominant role in the value of some environmental goods. These values are often understood as altruistically motivated. In terms of individual welfare, they cannot exist if people do not know about the goods. But, again, preferences do not reflect welfare only, but also views about other ends in addition to welfare. Hence, even if we knew for sure that no one would derive any welfare outside the survey context from the preservation of a species, it may still be worthwhile doing so. Many people may hold that the species by itself, or as a part of a larger ecosystem, is intrinsically valuable irrespective of human welfare. Valuation studies should therefore be designed to elicit such information.

Discussion and Conclusion

This chapter argued that when stated or revealed preferences do not reflect the maximization of individual welfare, it is welfare, rather than preferences, that should be pursued. This was illustrated with examples of imperfect information, limited cognitive capacity, and cognitive dissonance. It was also pointed out that the scientific community may suffer from shortsightedness and an inability to systematically deal with problems characterized by a small probability of a catastrophic outcome, and that policy-makers should bear this in mind. Also, when people's responses to stated preference surveys do not correspond to welfare maximization, this may indicate a nonanthropocentric view according to which animals and nature should be valued intrinsically, irrespective of humans' derived well-being. It is argued that public policy should reflect such views, but that more research is needed.

To illustrate these points, consider the following example. A referendum-mimicking survey is conducted to measure people's WTP for preserving a species. The estimated total WTP exceeds total cost (say 1 million US dollars) by 50 percent. No direct use values are involved, and currently living people who do not respond to the survey would not obtain any welfare improvements at all as a result of preserving the species. Should the species be preserved? Assuming that we have elicited the preferences correctly, the total WTP is larger than the costs. On the other hand, the fraction of the total population who responded to the survey is small, so the aggregate *welfare* effect of preserving the species is most likely smaller in monetary terms than the costs.[13] Yet people may have other motives in addition to the maximization of their own well-being. First, they may believe that future generations would be directly affected, and that their welfare should also count in today's decision-making. This is a legitimate motive and the WTP figures conveying such motives should not be excluded from consideration. Second, some respondents may think that the species should count irrespective of effects on human well-being. Thus, figures on individual

---

[13]   If 1 percent of the population answered the survey and their responses reflect the welfare effects for them, we have that these welfare effects are just 1.5 percent of the costs, or US $15,000. Recall that the welfare effects for others are zero.

WTPs, together with costs, are insufficient information for policy. We also need information on individual motives, ethical views, and cognitive strategies and limitations.

However, a caveat must be provided. Excessive paternalism, when policy-makers believe that people are making bad or irrational choices, may have negative *instrumental* effects. Indeed, we have seen enough terrible consequences of excessive paternalistic decision-making. Therefore, it may sometimes be advisable to "respect" individual preferences, rather than trying to determine the "true" individual welfare functions. This point parallels the recommendation made by Sidgwick (1874) and other utilitarians *against* the use of utilitarianism as a decision-making rule in daily life. The reason is that to adopt such a rule is in itself an action, the consequences of which may be good or bad. If one believes that the overall social consequences would be better if people instead applied some simplified agreed-upon ethical rules, then a convinced utilitarian would argue in favor of using these rules instead.[14] There is nothing inconsistent with this. Still, in our case it is also clear that people do often want to have restricted choices. Rational people know that they sometimes (or even often) make bad decisions, and they are aware of their limited self-control and cognitive capacity. Furthermore, it seems clear that all governments – including those that declare themselves to be liberal – apply a certain degree of paternalism. It is thus not reasonable to neglect analyzing deviations from respecting individual preferences. But, by the same token, it is important to bear instrumental considerations in mind when discussing policy recommendations in practice.

At the end of the day, how can the conclusions of this chapter be used? Can they be operationalized in a practically useful way, or do they only produce confusion and an increased (and possibly depressing) awareness that reality is awfully complex? Some of the conclusions are straightforward to apply. For example, if one knows that people's expressed preferences for a good are based on erroneous information, it may be possible to adjust a social BCA in order to better reflect actual individual welfare. Furthermore, it is possible, although more difficult, to elicit people's views about ethical ends in addition to individual well-being. Widely shared concerns for the environment or animal well-being *per se* should be reflected in public policy. However, the extent to which people hold such values, and the intensity of these values, are still uncertain and more research is needed.

Some conclusions are less straightforward. It does not appear practicable to try to measure degrees of cognitive dissonance, risk misperception, and short-sightedness on individual basis in order to construct a modified BCA. On the

---

[14] This is the standard utilitarian response to the popular example about killing a healthy person who is walking outside a hospital in order to save two dying persons, one in desperate need of a new heart and the other of a new liver. Everything else being equal, this would probably be good according to utilitarian ethics. However, everything else would not be equal! It is easy to imagine the far-reaching consequences of the fact that people could not walk safely outside a hospital, or even elsewhere. Therefore, utilitarians would agree with nonutilitarians that it would certainly be a very bad idea to kill the healthy person.

other hand, many economic relations and variables, such as labor supply and human capital relations, are difficult to quantify but still useful for decision-making. An insight can thus be practically useful even when it is difficult to quantify, and when it cannot be added to other information in order to construct a unique index of goodness.[15] So, even though we may not be able to measure cognitive dissonance, or the difference between individual welfare and preferences, very accurately (and most often do not measure them at all), the insights of such information may still be of practical importance in the decision-making process. In all public policy-making there is a certain amount of quantified information, as well as more qualitative information. This chapter hopefully contributes to the appropriate use of the quantified information, such as estimated benefits and costs, and to understanding their limitations. In particular, it offers arguments on why deviations from the conventional BC rule may be socially preferable and, perhaps even more importantly, in what direction such a deviation should go in different cases.

Finally, it should be emphasized that the mere existence of a deviation from the standard assumptions of economics should *not* be seen as an argument that supports an "anything goes" alternative. To influence a political decision through, for example, a cognitive dissonance argument, some indication must be provided of its importance in the particular case, both qualitatively and quantitatively.

References

Akerlof, G. 1991: Procrastination and obedience. *American Economic Review*, 81(2), 1–19.

Akerlof, G. and Dickens, T. W. 1982: The economic consequences of cognitive dissonance. *American Economic Review*, 72, 307–19.

Andreoni, J. 1990: Impure altruism and donations to public goods: a theory of warm glow giving. *Economic Journal*, 100, 464–77.

Baxter, W. F. 1974: *People or Penguins: The Case for Optimal Pollution*. New York: Columbia University Press.

Beattie, J., Covey, J., Dolan, P., Hopkins, L., Jones-Lee, M., Loomes, G., Pidgeon, N., Robinson, A., and Spencer, A. 1998: On the contingent valuation of safety and the safety of contingent valuation: part 1 – *caveat investigator*. *Journal of Risk and Uncertainty*, 17(1), 5–26.

Becker, G. S. 1993: Nobel Lecture: The economic way of looking at behavior. *Journal of Political Economy*, 101(3), 385–409.

Broome, J. 1991a: Utility. *Economics and Philosophy*, 7, 1–12.

Broome, J. 1991b: *Weighing Goods*. Cambridge, MA: Blackwell.

Broome, J. 1999: *Ethics Out of Economics*. Cambridge: Cambridge University Press.

Brown, G. and Layton, D. F. 1996: Resistance economics: social cost and the evolution of antibiotic resistance. *Environment and Development Economics*, 13, 349–55.

---

[15]   To give an illustration from another field, consider the (difficult) art of bringing up children. This is an area where an enormous amount of advice exists. Clearly, some advice can be practically useful, even if it is not used as an input, to construct a one-dimensional index to measure whether a certain action is good or bad.

Carthy, T., Chilton, S., Covey, J., Hopkins, L., Jones-Lee, M., Loomes, G., Pidgeon, N., and Spencer, A. 1998: On the contingent valuation of safety and the safety of contingent valuation: part 2 – the CV/SG "chained" approach. *Journal of Risk and Uncertainty*, 17(3), 187–214.

Conlisk, J. 1996: Why bounded rationality. *Journal of Economic Literature*, 34(2), 669–700.

Doessel, D. P. 1998: The "sleeper" issue in medicine: Clem Tisdell's academic scribbling on the economics of antibiotic resistance. *International Journal of Social Economics*, 25(6–8), 956–67.

Festinger, L. 1957: *A Theory of Cognitive Dissonance*. Stanford, CA: Stanford University Press.

Graaf, J. de V. 1957: *Theoretical Welfare Economics*. Cambridge: Cambridge University Press.

Gregory, R. and Slovic, P. 1997: A constructive approach to environmental valuation. *Ecological Economics*, 21(3), 175–81.

Gregory, R., Lichtenstein, S., and Slovic, P. 1993: Valuing environmental resources: a constructive approach. *Journal of Risk and Uncertainty*, 72, 177–97.

Harrison, G. W. 1992: Valuing public goods with the contingent valuation method: a critique of Kahnemann and Knetsch. *Journal of Environmental Economics and Management*, 23(3), 248–57.

Harrod, R. F. 1948: *Towards a Dynamic Economics*. London: St. Martin's Press.

Harsanyi, J. C. 1982: Morality and the theory of rational behavior. In A. K. Sen and B. Williams (eds.), *Utilitarianism and Beyond*. Cambridge: Cambridge University Press.

Harsanyi, J. C. 1995: A theory of prudential values and a rule utilitarian theory of morality. *Social Choice and Welfare*, 12(4), 319–33.

Harsanyi, J. C. 1997: Utilities, preferences, and substantive goods. *Social Choice and Welfare*, 14(1), 129–45.

Hausman, J. A. 1979: Individual discount rates and the purchase and utilization of energy-using durables. *Bell Journal of Economics*, 10(1), 33–54.

Johansson-Stenman, O. 1998: The importance of ethics in environmental economics with a focus on existence values. *Environmental and Resource Economics*, 11(3–4), 429–42.

Johansson-Stenman, O. 2000: Should policy makers be concerned with the subjective or objective risk, or both? Working paper, Department of Economics, Göteborg University.

Jones-Lee, M. and Loomes, G. 1997: Valuing health and safety: some economic and psychological issues. In R. Nau (ed.), *Economic and Environmental Risk and Uncertainty*. Amsterdam: Kluwer, 3–32.

Kahneman, D. and Knetsch, J. L. 1992: Valuing public goods: the purchase of moral satisfaction. *Journal of Environmental Economics and Management*, 22(1), 57–70.

Kahneman, D. and Tversky, A. 1979: Prospect theory: an analysis of decision under risk. *Econometrica*, 47, 263–91.

Marglin, S. 1963: The social rate of discount and the optimal rate of investment. *Quarterly Journal of Economics*, 77, 95–111.

Mitchell, R. C. and Carson, R. T. 1989: *Using Surveys to Value Public Goods: The Contingent Valuation Method*. Washington, DC: Resources for the Future.

Ng, Y.-K. 1999: Utility, informed preference, or happiness: following Harsanyi's argument to its logical conclusion. *Social Choice and Welfare*, 16(2), 197–216.

O'Neill, O. 1997: Environmental values, anthropocentrism and speciesism. *Environmental Values*, 6(2), 127–42.

Pigou, A. C. 1929: *The Economics of Welfare*. London: Macmillan.

Pollak, R. A. 1998: Imagined risks and cost–benefit analysis. *American Economic Review*, 88(2), 376–80.

Ramsey, F. 1928: A mathematical theory of savings. *Economic Journal*, 138, 543–59.

Russell, C. S., Bjømer, T. B., and Clark, C. D. forthcoming: Searching for evidence of alternative preferences, public as opposed to private. *Journal of Economic Behavior and Organization.*

Samuelson, P. A. 1938: A note on the pure theory of consumer behaviour. *Econometrica*, 5, 61–71.

Samuelson, P. A. 1947: *Foundations of Economic Analysis.* Cambridge, MA: Harvard University Press.

Samuelson, W. and Zeckhauser, R. 1988: Status quo bias in decision making. *Journal of Risk and Uncertainty*, 1(1), 7–59.

Schkade, D. A., and Payne, J. W. 1994: How people respond to contingent valuation questions: a verbal protocol analysis of willingness to pay for an environmental regulation. *Journal of Environmental Economics and Management*, 26, 88–109.

Sen, A. K. 1977: Rational fools: a critique of the behavioural assumptions of economic theory. *Philosophy and Public Affairs*, 6, 317–44.

Sen, A. K. 1979: Personal utilities and public judgements: or what's wrong with welfare economics? *Economic Journal*, 89, 537–58.

Sen, A. K. 1985: Goals, commitment, and identity. *Journal of Law, Economics and Organization*, 1(2), 341–55.

Sen, A. K. 1987: *On Ethics and Economics.* Oxford: Blackwell.

Sen, A. K. 1991: Utility – ideas and terminology. *Economics and Philosophy*, 7, 277–83.

Sidgwick, H. 1874: *The Methods of Ethics.* London: Macmillan.

Singer, P. 1975: *Animal Liberation: A New Ethics for Our Treatment of Animals.* New York: Avon.

Singer, P. 1979: *Practical Ethics.* Cambridge: Cambridge University Press.

Slovic, P. 1995: The construction of preference. *American Psychologist*, 50(5), 364–71.

Slovic, P. 2000: *The Perception of Risk.* London: Earthscan.

Slovic, P. and Lichtenstein, S. 1983: Preference reversals: a broader perspective. *American Economic Review*, 73(4), 596–605.

Spash, C. L. and Hanley, N. 1995: Preferences, information and biodiversity preservation. *Ecological Economics*, 12, 195–208.

Stephan, P. 1996: The economics of science. *Journal of Economic Literature*, 34(3), 1199–235.

Stevens, T. H., Echeverria, J., Glass, R. J., Hager, T., and Moore, T. A. 1991: Measuring the existence value of wildlife: What do CVM estimates really show? *Land Economics*, 67(4), 390–400.

Thaler, R. H. 1992: *The Winner's Curse: Paradoxes and Anomalies of Economic Life.* Princeton, NJ: Princeton University Press.

Thaler, R. H. 2000: From *Homo economicus* to *Homo sapiens. Journal of Economic Perspectives*, 14(1), 133–41.

Tversky, A. and Kahneman, D. 1974: Judgements under uncertainty: heuristics and biases. *Science*, 185, 1124–31.

Tversky, A., Slovic, P., and Kahneman, D. 1990: The causes of preference reversal. *American Economic Review*, 80(1), 204–17.

Vatn, A. and Bromley, D. W. 1994: Choices without prices without apologies. *Journal of Environmental Economics and Management*, 26(2), 129–48.

Viscusi, W. K. 1992: *Fatal Tradeoffs, Public and Private Responsibilities for Risk.* Oxford: Oxford University Press.

Viscusi, W. K. 1998: *Rational Risk Policy.* Oxford: Oxford University Press.

CHAPTER **8**

Awkward Choices: Economics and Nature Conservation

Nick Hanley and Jason F. Shogren[1]

In this chapter, we look at the reasons why economists believe that decisions reached over nature conservation using economic analysis are in some sense better than decisions reached without such analysis. We then consider three types of objection to this view, relating to rights, uncertainty, and futurity. However, we find that a powerful case (and in some senses, an inevitable case) can still be made for using economics in decision-making over nature conservation, although how exactly this is done is open to debate.

The Case for Economics

The overwhelming view amongst conventional economists is that decisions made using their tools are likely to be superior to decisions made that ignore these tools (Shogren et al., 1999). While economics cannot be the only discipline to behave in this way, it is worthwhile reviewing the arguments put forward to support this view, using the example of nature conservation. Decisions about conservation involve reallocations of resources. For instance, deciding to conserve landscape and wildlife habitats on an island rather than allow development of an immense quarry imply a diversion of resources away from mining activities and toward the production of environmental public goods. How can we decide whether such a redistribution is desirable for society? Economists answer with the Potential Pareto Improvement (PPI) criterion: Can the gainers compensate the losers and still be better off? In this case, are the benefits of preserved landscape and wildlife higher than the losses of forgone quarry products output? This way of judging outcomes, which involves the summing up of money equivalents of welfare changes across individuals ("welfarism") is argued to be superior to, for instance, referenda, as it takes into account both the direction and intensity of preferences. What is more, economists claim to be able to rank

[1] Thanks go to Dan Bromley, Jouni Paavola, and Neil Summerton for helpful comments on an earlier version.

competing uses of scarce resources on the basis of this comparison of benefits and costs.

The net present value measure derived from the empirical application of welfarism, namely benefit–cost analysis, is claimed to be able to produce a ranking of competing resource allocations (for example, in our case, turning the island into a new Disney World, rather than either of the two alternatives so far discussed), although many caveats are attached to this property (see Johansson, 2000). The balancing of benefits and costs seems, moreover, an eminently sensible way of judging alternatives, and is something that we observe people implicitly doing everyday (although, of course, this is not the only way decisions are made). As Randall (2000) has observed, it is hard to argue against giving some consideration to preferences in social decision-making. Recent advances in environmental valuation (Garrod and Willis, 1998) make it possible to include environmental impacts in the welfarist approach, as enacted through benefit–cost analysis.

Economists also claim that economics can help in designing policies that achieve targets more efficiently, irrespective of how these targets are determined. This retains a role for economics in environmental policy-making even if we reject the welfarist approach to setting targets. The crucial insight is to recognize that the costs of achieving targets vary across those people/firms whose behavior we wish to change. Thus, since Baumol and Oates (1971) and Montgomery (1972), it has been known that economic incentives such as tradable permits or pollution taxes can achieve target reductions in pollution at lower social cost than regulatory alternatives. Given that resources are scarce, economists argue, we should be interested in this efficiency property, since it recognizes the opportunity costs of environmental management (for example, the impacts of needlessly costly regulation on economic growth).[2] If a regional conservation agency needs to reduce pollution inputs to an estuary in order to conserve fish stocks, then it is likely to be able to do this more cheaply if it makes at least some use of economic instruments, rather than relying entirely on a command-and-control approach.

More generally, it makes sense to recognize the varying marginal costs of achieving environmental targets, and a large literature exists that compares the resource costs of alternative means to achieve conservation objectives. This includes comparisons of land purchase versus management agreements in the English Norfolk Broads (Colman, 1991); alternative strategies for protecting a Safe Minimum Standard population for woodpeckers in the US (Hyde, 1989); alternative approaches to native pinewoods regeneration in the Scottish Highlands (MacMillan, Harley and Morrison, 1999); and of policies for reducing grazing pressure in heather moorlands in Shetland (Hanley et al., 1998). Ando et al. (1998), for example, have pointed out that if a conservationist accounted for the different prices of land across the US, he or she could reduce by two-thirds the costs of protecting half the species on the Endangered Species Act list. Ecological complexities do, however, complicate the use of economic instruments to achieve

---

[2] There is actually considerable evidence that regulators place a much lower weight on the efficiency criterion than economists might assume (see Hanley, Moffatt, and Hallett, 1990).

conservation objectives; for example, in the context of red deer management in Scotland (MacMillan, 2000).

Third, economists claim that economics can help make policy more effective by acknowledging the opportunity costs that land managers face in playing their part in nature conservation. Failure to recognize these costs, and the incentives that they create, can lead to ineffective and self-defeating policy. Two examples illustrate this. Most biodiversity in the UK is found on private land. The state must therefore recognize the opportunity costs to land managers (typically farmers) of protecting this diversity. If a farmer is to be persuaded not to drain a wetland, or not to increase stocking rates, some financial compensation is necessary. Setting payment rates below marginal opportunity costs will result in too low a level of participation. What is more, government intervention to support farm incomes (for example, by guaranteeing prices) increases this opportunity cost above the level that the market would set, thus increasing the cost of conservation. Hanley, Whitby, and Simpson (1999) report evidence of the link between high commodity support payments and low sign-up rates in several agri-environmental schemes in the UK.

Incentives also matter to participation and the cost of conservation in the case of bilaterally negotiated contracts, as is the case for management agreements under the Wildlife and Countryside Act 1981 (Spash and Simpson, 1994). Under such contracts, farmers have incentives to threaten actions to deplete biodiversity even if these would be unprofitable, since by doing so they may be able to extract information rents in the form of payments from conservation agencies. The high costs and low effectiveness of bilateral negotiation schemes was a major reason for the change to fixed-rate voluntary participation schemes under current agri-environmental policy in the UK.

Incentive structures may also lead to landowners seeking to disguise the ecological qualities of their land, fearing that discovery may lead to "conservation blight" (a reduction in the value of their land). For this reason, policy designs that reward landowners for keeping ecological quality high may be more effective than regulatory alternatives.

Objections to Economics

There are many objections to the three uses of economics suggested above. Perhaps the greatest relate to welfarism, and its empirical form, benefit–cost analysis. In this section, we review three areas of concern, which are referred to here as rights and tradeoffs, uncertainty, and futurity.[3]

Rights and tradeoffs

One fairly well-rehearsed criticism of benefit–cost analysis (BCA) in general, and of environmental valuation in particular, is that it does not represent the way in

---

[3] This means, of course, that we omit other objections to BCA, such as those based around income distribution issues.

which at least some people view the environment. BCA is based on the concept of tradeoffs: a benefit for a cost, a loss for a gain. As long as losses are less than gains, then those who lose out could potentially be compensated by those who gain, and a net gain could still remain. This compensation for losses is at the heart of the PPI, as several authors have pointed out (see Griffin, 1995; Farrow, 1998). However, the PPI criterion assumes that all losses are, in principle, compensatable, and that losses are essentially identical to negative gains (that is, are commensurate). It also assumes that all that matters are end-points – namely, *how much* net benefit society gets from a resource allocation, rather than *how it gets it*.

A related issue concerns the economic theory underlying environmental valuation methods, which treats the environment no differently to any other commodity (see, for instance, Braden and Kolstad, 1991). How we represent theoretically the welfare gains of having cleaner rivers is identical to how we represent the welfare gains from having more mobile phones. This is because the welfare measures used to monetize both come from the same comparison of before-and-after utility: How much income would I give up to have more of something I like, such as wildlife preservation (maximum willingness to pay, WTP); or how much of an increase in income would I have to be offered to tolerate an increase in the level of something I do not like, such as pollution (my minimum willingness to accept, WTA)?

Authors such as Spash and Hanley (1995) and Stevens et al. (1991) present empirical evidence that shows that this treatment of the environment does not fit in with some peoples' views. First, they claim that a significant portion of the population has a rights-based attitude to environmental resources such as wildlife: such individuals feel that wildlife has the right to be protected, regardless of its utility value. Second, some individuals refuse to contemplate *any* tradeoff of their income for changes in environmental quality. This rejection of the concept of tradeoffs occurs typically for prospective environmental losses (such people would demand infinite compensation for environmental losses), but also to a degree for environmental gains. These views are often characterized as "lexicographic preferences." Philosophers such as Holland (1995) and Sagoff (1988) have noted that economists should not be surprised by these results, since their underlying model (utilitarianism) is a poor description of common ethical positions toward the environment. The ultimate conclusion is that BCA should not guide our actions over the environment.

However, some objections can be raised to this view. First, individuals who claim a belief in rights for the environment may change their minds when the costs to them of protecting these rights becomes apparent. Results from a survey undertaken by Hanley and Milne (1996), concerned with protecting wildlife and landscape resources in Scotland, are interesting in this regard (table 8.1). As may be seen, almost all (99 percent) of respondents thought that wildlife and landscape have the right to be protected. When it was pointed out that this "costs money or costs jobs," this reduced the percentage of those saying "yes" to 49 percent, a minority. When asked to imagine that protection cost the specific sum of 10 percent of their income, the "yes" vote fell further, to 38 percent of the

Table 8.1 The percentage of the sample supporting environmental rights: private income

Question	Proportion saying "yes" out of total sample (%)
Do wildlife and landscape have the right to be protected?	99
Even when this costs jobs/money?	49
If the cost was 10% of your income?	38
If the cost was 25% of your income?	19

total; and when asked to imagine that protection cost 25 percent of their income, this fell further still, to 19 percent. The implication is that people's views in this sample regarding the "rights" of wildlife and landscape to be protected are negatively related to the cost of protecting these rights. Almost one-fifth of the sample, however, still claim to maintain these rights-based views even when 25 percent of their income is in (hypothetical) prospect of being lost. The view of "modified" lexicographic preferences put forward by Spash and Hanley suggests that persons who believe in environmental rights can still be willing to pay to stop a decrease in environmental quality, but that the amount should be invariant with the size of environmental loss in prospect. Hanley and Milne found that belief in rights was positively correlated with WTP, so that people with rights-based beliefs were likely to offer higher bids to protect the environment than those without such beliefs (see also Spash, ch. 13, this volume). This seems sensible: if the only game in town is that environmental protection must be paid for, then those who care most about it offer, *ceteris paribus*, the highest bids.

Finally, Sagoff (1988) has offered the view that the focus of economists on the individual as a consumer of environmental quality is amiss, since people are more likely to think about environmental issues from the perspective of citizens. This would imply them being less concerned about the implications of resource allocations for their own welfare, and more concerned with implications for the welfare of society as a whole (note that this is still compatible with welfarism, but is incompatible with the standard economic model of environmental valuation). Hanley and Milne tried to discover whether reframing the issue of potential compensation in terms of community well-being would make much difference to rights-based objections to the PPI criterion (table 8.2). As may be seen, the most noticeable changes are in WTP situations. However, overall, reframing the question in terms of community incomes does not seem to make much difference in practice according to this evidence.

It is undoubtedly true that some people reject economists' conception of environmental values, and the use of BCA in an environmental context. To describe such people as "lexicographic" may or may not be useful. However, to proceed from the former observation to rejecting BCA and welfarism for environmental decision-making is problematic for several reasons. First, while people may refuse tradeoffs for some types of environmental loss, they may not refuse them for others. Second, when environmental gains are in prospect,

Table 8.2 The percentage of the sample supporting environmental rights: private versus collective income

Response	Exchange higher personal income for lower environmental quality?	Exchange lower personal income for higher environmental quality?	Exchange higher community income for lower environmental quality?	Exchange lower community income for higher environmental quality?
Yes, definitely	32	8	33	9
Yes, probably	50	32	46	16
No, probably	16	40	17	53
No, definitely	1	18	4	22

ethical attitudes at odds with the utilitarian seem less threatening to BCA, since those with rights-based beliefs seem willing to play the economic game. Third, rights-based views can be seen as being at odds with social democracy: if 99 people benefit from a project to drain a swamp, but one person feels worse off, what do we make of this single loser's insistence that his or her infinite WTA should rule the day?

Uncertainty and ignorance

Two issues are relevant here. The first concerns the inability of BCA to make much sense of true uncertainty. The second refers to problems of ignorance on the part of consumers.

BCA does not cope well with true uncertainty. Consider the example of the environmental costs of a new pollutant entering a river. Three situations regarding our knowledge of these environmental costs are possible. First, scientists may be unsure about what physical impacts the pollutant will have; this implies that not all "states of the world" $s_1, \ldots, s_n$ are known. Second, scientists may be able to identify all possible states $s_1, \ldots, s_n$ but may not be able to identify their probability distribution. Third, all possible states of the world and their probability distribution may be known. Most treatments of *risk* in economics are concerned with the circumstances of the third case, but not of the first two. If we know all possible states of the world and their probabilities, then expected values can be estimated along with their certainty equivalents. These can then form part of a BCA. However, if not all states of the world are known, or if their probabilities are unknown, then we face a situation of true uncertainty. In this case, which is likely to describe many environmental management situations, BCA must then fall back on sensitivity analysis, which estimates net benefits under different states of the world. We could then base decision-making, if desired, on tools from decision science, such as the minimax regret criterion.

One policy area that is characterized by extreme uncertainty is climate change. Economists are divided on the best response to such uncertainty. As Tietenberg

(1998) points out, we do not even know the shape of the damages curve, which may be highly nonlinear. The costs of control vary according to how greenhouse gas emissions are reduced, and in terms of who undertakes these reductions (which countries, which agents). Given the huge uncertainties involved, Tietenberg argues, it is amiss to set targets for climate change policy using BCA. Pearce (1998), however, argues that by *not* using BCA to set targets, we risk making a very expensive mistake. The fact that uncertainty is high is a reason for doing BCA, rather than following either of two alternative decision rules: (i) act now because of the precautionary principle; or (ii) do nothing now until we learn more. Pearce notes that estimates do exist of control costs and avoided damages, and that the desired direction of change in emissions can be determined from an examination of these relative values. Pearce calls for a BCA that "embodies these uncertainties." He also argues that BCA should be done because it identifies *who* gains and loses from climate change and by how much: for example, developed countries lose 1.3–1.6 percent of GDP per annum, while developing countries lose 1.6–2.7 percent of GDP in his analysis. This shows, according to Pearce, that developing countries should not be exempt from taking partial responsibility for reducing GHG emissions. In contrast, Tietenberg states that deciding not to reduce GHG emissions on BCA grounds would be unwise even if currently estimated marginal costs of control are greater than currently estimated marginal benefits, since future damages are *potentially* enormous and are also irreversible.

A second objection to the use of economic principles in environmental decision-making concerns the level of information that consumers hold. Welfare economics is based on the presumption that consumers know what is best for them, while the PPI rule says that social welfare depends solely on the utilities of these consumers. This may be all well and good for decisions that involve familiar resources: we can be trusted to make the best decision as to whether eating carrots or turnips for dinner gives us greater welfare. But, ecologists argue, how can we base decision-making over complex environmental issues on the preferences of poorly informed consumers? If the decision is over whether to preserve a bog, or whether to develop it into a theme park, how can the money measures of environmental cost (derived from willingness to pay, and thus from preferences) adequately reflect the ecological importance of the bog, which is hard even for specialists to understand? And if we use BCA to decide which species to target protection toward, will the use of economic valuation methods not bias protection in favor of "warm and cuddly" species rather than snakes and insects?

It is certainly true that many consumers have a poor understanding of some environmental issues – for example, biodiversity (for evidence, see Spash and Hanley, 1995). It is also true that telling people more about these environmental issues changes their stated preferences (Munro and Hanley, 1999). This raises the awkward issue of how much information is "enough" on which to base social decision-making. In addition, alternative decision-aiding methodologies such as citizen juries can be better ways of informing ordinary people about complex environmental issues. Citizen juries do not, however, produce outputs that are commensurate with the PPI criterion, although finding ways of combining the best features of juries and environmental valuation methods is an important avenue for future research. However, the fact of citizen ignorance should not deter us from using the PPI criterion. Alternative valuation methods, such as ecosystem service methods, can in some cases avoid the ignorance problem, since they avoid the direct questioning of respondents. People can be also led into grasping the important basics of environmental issues in the context of a contingent valuation survey. While this probably means that preferences for those in the sample may now differ from those in the general public (a manifestation of the much more general phenomenon that by measuring something we change it), this does not mean that informed environmental values are a poor guide to policy-making. The alternative, after all, is to bias decision-making toward the views of experts, and to neglect the views of the general public. While this would no doubt appeal to many experts, it is hardly democratic.

Futurity

Decisions made now over nature conservation impinge on the future. For instance, a decision as to whether or not to go ahead with a reintroduction program for the European beaver in Scotland has implications not just for people alive now, who may enjoy viewing beavers (or for those – such as farmers and fishermen – who may be opposed to their introduction), but also for future people who might have benefited or lost out from a reintroduction program.[4] In applying BCA to such a decision today, all we have to go on are the values of those people currently alive, in terms of their WTP to either have or stop the program. While present-generation WTP bids include future values to some extent (nonuse values may be motivated by bequest motives), they seem likely to be a poor measure of values for those in the future. As Shogren (1999) has pointed out, this is for two reasons. First, we do not know the preferences of future individuals. Will they be more or less pro-beaver? Second, we do not know what the opportunity sets facing future people will look like. Will biodiversity be greater or less than now? What will be the relevant relative prices that these future people face? Our only guide to both these questions is current observations.

Another example relates to climate change impacts on recreational fishing in the US (Ahn et al., 2000). The fact that we can estimate current consumer surplus values and how they are predicted to change under currently forecasted changes in regional precipitation and temperature may give us a false sense of security in terms of what these welfare changes actually turn out to be. Climate change might affect household resources, human resource investment prices and levels, endowments, preferences, labor market opportunities, and the natural environment that influences our descendants' opportunity sets – the basic material needed for attainment in life. Climate change risks indirectly modify our heirs' life chances by reducing and reallocating household resources or by constraining their choices, or both. Our descendents may shift resources toward a

---

[4] It may seem odd to talk about people "losing out" from species reintroductions, but this is precisely what MacMillan, Duff, and Elston (forthcoming) found for the wolf in Scotland.

sick child and away from recreation. Their children might then have to forgo the experience of fishing the same river as their ancestors (Shogren, 1999).

In terms of preferences, the best guess we have about future people is that they will be pretty much like us. The key is probably to make better predictions of changes in opportunity sets, and this is something that economists can certainly not do alone.

Awkward Choices: Why Economics is Still Needed

Economists often repeat the mantra that "resources are scarce." This is because it is a fact, both at the global and local levels. The fact that resources such as land are scarce means that deciding to use land for one purpose (nature conservation) means we forgo a return from using it in some other way (intensive farming). The fact that governments have limited scope for public spending means that each extra pound spent on nature conservation objectives is one less pound to spend on hospitals or schools. The fact that nature conservation agencies operate on limited budgets means that they must effectively prioritize species action plans when it comes to protecting biodiversity. Scarcity forces awkward choices upon us, and economics offers us the means of both recognizing the resultant opportunity costs; and of weighing up the advantages and disadvantages of alternative allocations of these scarce resources. The PPI criterion is, in a way, the taking to extremes of the weighing up approach, since it claims to be able to decide which allocation from amongst many best serves social welfare. Perhaps one positive action that economists could take in the face of the kinds of criticisms of welfarism outlined above (and elsewhere in this book) is to point out that application of the PPI criterion is only one way in which BCA can be used. BCA can also be used as a means of making tradeoffs clear, and of laying out the good and bad points of a project/policy in a systematic way. This alternative role for BCA does not even need to extend to full monetization of all impacts, which is another aspect of BCA that worries many (Hanley, 2001).

However, it is worthwhile inquiring whether the welfarist function of BCA can be retained in the face of the kinds of criticisms set out in the previous section. In terms of ethical objections, we might say – as Randall (2000) has suggested – that the main desire is to rule certain environmental changes as being "off-limits" to BCA. The key problem, of course, is knowing how to define these limits. As the extent of limits rises, so does their shadow price in opportunity cost terms. In terms of futurity, the main concern appears to be that we should avoid doing things that impose undue burdens on future generations. The intergenerational equity impacts of discounting are well known (Lind, 1982), but what is being discussed here is an additional concern: namely, that we know too little of the preferences and opportunity sets of future generations to make good decisions on their behalf. This is at the heart of economic conceptions of sustainability, and our best response is probably to be found in this literature, for instance in terms of enforcing/encouraging a positive rate of genuine savings or imposing a nondeclining natural capital stock rule. Finally, uncertainty and ignorance characterize many aspects of environmental decision-making. Again, the use of BCA subject to constraints appears to be the appropriate response here; for example, in terms of safe minimum standards (Randall and Farmer, 1995).

None but the most manic economist would insist that BCA and the PPI criterion are *all* that we need to pay attention to in making decisions about anything, still less nature conservation. However, our view is that in a world of awkward choices, the information contained within good economic analysis is much too useful to disregard.

References

Ahn, S., de Steiguer, J., Palmquist, R., and Holmes, T. 2000: Economic analysis of the potential impact of climate change on recreational trout fishing in the southern Appalachian mountains: an application of a nested multinomial logit model. *Climatic Change*, 45, 493–509.

Ando, A., Camm, J., Polasky, S., and Solow, A. 1998: Species distributions, land values, and efficient conservation. *Science*, 279, 2126–8.

Baumol, W. J. and Oates, W. E. 1971: The use of standards and prices for the protection of the environment. *Swedish Journal of Economics*, 73, 42–54.

Braden, J. and Kolstad, C. 1991: *Measuring the Demand for Environmental Quality*. Amsterdam: North Holland.

Colman, D. 1991: Land purchase as a means of providing external benefits from agriculture. In N. Hanley (ed.), *Farming and the Countryside: an Economic Analysis of External Costs and Benefits*. Wallingford, UK: CAB International, 215–29.

Farrow, S. 1998: Environmental equity and sustainability: rejecting the Kaldor–Hicks criterion. *Ecological Economics*, 27, 183–8.

Garrod, G. and Willis, K. 1998: *Economic Valuation and the Environment*. Cheltenham, UK: Edward Elgar.

Griffin, R. 1995: On the meaning of economic efficiency in policy analysis. *Land Economics*, 71, 1–15.

Hanley, N. 2001: Cost–benefit analysis and environmental policy-making in the UK. *Environment and Planning C: Government and Policy*, 19, 103–18.

Hanley, N. and Milne, J. 1996: Ethical beliefs and behaviour in contingent valuation. *Journal of Environmental Planning and Management*, 39, 255–72.

Hanley, N. D., Moffat, I., and Hallett, S. 1990: Why is more notice not taken of economists' prescriptions for the control of pollution? *Environment and Planning A*, 22, 1421–39.

Hanley N., Whitby, M., and Simpson, I. 1999: Assessing the success of agri-environmental policy in the UK. *Land Use Policy*, 16, 67–80.

Hanley, N., Kirkpatrick, H., Oglethorpe, D., and Simpson, I. 1998: Paying for public goods from agriculture: an application of the Provider Gets Principle to moorland conservation in Shetland. *Land Economics*, 74, 102–13.

Holland, A. 1995: The assumptions of cost–benefit analysis: a philosopher's view. In K. Willis and J. Corkindale (eds.), *Environmental Valuation: New Perspectives*. Wallingford, UK: CAB International, 21–38.

Hyde, W. 1989: Marginal costs of managing endangered species: the red cockaded woodpecker. *Journal of Agricultural Economics Research*, 41(2), 12–20.

Johansson, P.-O. 2000: Microeconomics of valuation. In H. Folmer and H. Landis Gabel (eds.), *Principles of Environmental and Resource Economics*, 2nd edn. Cheltenham, UK: Edward Elgar, 34–71.

Lind, R. C. 1982: *Discounting for Time and Risk in Energy Policy*. Baltimore, MD: Johns Hopkins University Press.

MacMillan, D. 2000: An economic case for land reform. *Land Use Policy*, 17, 49–57.

MacMillan, D., Duff, E., and Elston, D. forthcoming: Modelling the non-market environmental costs and benefits of biodiversity projects using contingent valuation data. *Environmental and Resource Economics*.

MacMillan, D., Harley, D. and Morrison, R. 1999: Cost-effectiveness analysis of woodland ecosystem restoration. In M. O'Connor and C. Spash (eds.), *Valuation and the Environment: Principles and Practices*. Cheltenham, UK: Edward Elgar, 109–22.

Montgomery, W. D. 1972: Markets in licenses and efficient pollution control. *Journal of Economic Theory*, 5, 395–418.

Munro, A. and Hanley, N. 1999: Information, uncertainty and contingent valuation. In I. J. Bateman and K. G. Willis (eds.), *Contingent Valuation of Environmental Preferences: Assessing Theory and Practice in the USA, Europe, and Developing Countries*. Oxford: Oxford University Press, 258–79.

Pearce, D. 1998: Economic development and climate change. *Environmental and Development Economics*, 3, 389–91.

Randall, A. 2000: Taking benefits and costs seriously. In H. Folmer and T. Tietenberg (eds.), *The International Yearbook of Environmental and Resource Economics 1999/2000*. Cheltenham, UK: Edward Elgar, 250–72.

Randall, A. and Farmer, M. 1995: Benefits, costs and the Safe Minimum Standard of conservation. In D. W. Bromley (ed.), *Handbook of Environmental Economics*. Oxford: Blackwell, 26–44.

Sagoff, M. 1988: *The Economy of the Earth*. Cambridge: Cambridge University Press.

Shogren, J. 1999: Speaking for citizens from the far future. *Climate Change*, 5, 489–91.

Shogren, J. F., Tschirhart, J., Anderson, T., Ando, A W., Beissinger, S. R., Brookshire, D., Brown, Jr., G. M., Coursey, D., Innes, R., Meyer, S. M., and Polasky, S. 1999: Why economics matters for endangered species protection. *Conservation Biology*, 13, 1257–67.

Spash, C. and Hanley, N. 1995: Preferences, information and biodiversity preservation. *Ecological Economics*, 12, 191–208.

Spash, C. and Simpson, I. 1994: Utilitarian and rights-based alternatives for protecting sites of special scientific interest. *Journal of Agricultural Economics*, 45, 15–26.

Stevens, T. H., Echeverria, J., Glass, R. J., Hager, T., and More, T. A. 1991: Measuring the existence value of wildlife: What do CVM estimates really show? *Land Economics*, 67, 390–400.

Tietenberg, T. 1998: Economic analysis and climate change. *Environment and Development Economics*, 3, 402–4.

PART IV

Ethical Dimensions of Policy Consequences

9

All Environmental Policy Instruments Require a Moral Choice as to Whose Interests Count

A. Allan Schmid[1]

Environmental policy requires tragic choices among the interests of different groups – tragic in the sense that frequently one must choose sides, because not everyone can win. This requires moral or ethical choice. The term "moral choice" is used here to denote a category of choice where one person's choices substantially affect others. It is not meant to distinguish moral from immoral, which is a value judgment.[2] The market with its promise of voluntary choice is often offered as a way to avoid moral choices. However, this ignores the prior public and moral choice of who is buyer and who is seller – of opportunities to use landscape resources, for example. All policy instruments are coercive if interests differ. All involve planning and collective action. All can be designed to be substitutes and complements in terms of shaping performance and all grapple with ubiquitous externalities. When you choose who is seller and who is buyer of opportunities, you decide the distribution of income, implement an ethical choice, and influence what kind of a world we live in.

In what follows, markets, regulation, courts, and public spending will be examined as examples of alternative policy instruments. The implications of these policy instruments will be illustrated in the context of landscape and countryside issues.

Markets

Markets are often advanced as a voluntary alternative to the so-called "command and control" of regulations. A popular phrase contrasts "command and control" with

[1] Thanks go to Dan Bromley, Sandra Batie, Warren Samuels, and Jouni Paavola for their critical review.
[2] There is an old debate in philosophy about the existence of intrinsic values, which is argued by some environmental philosophers. The argument in this chapter is that any policy requires a moral

voluntary choice. Another contrasts "coercive" regulations with "free" markets. This is mischievous, if not devious. At least, it is certainly selective perception. First of all, the market is not a single unique thing. There are as many markets as there are starting place ownership structures. I personally love markets, but of course I always want to be a seller of opportunities and not a buyer. Equally mischievous is the idea that externalities are a special case where markets may fail. Instead, externalities are the ubiquitous stuff of scarcity and interdependence.

Animal feeding pollution

Let's explore these arguments in greater detail in the context of large-scale concentrated animal feeding operations, which is one of the most contentious countryside issues in contemporary Europe and the United States. Operators of large animal feedlots are in conflict with their neighbors, who are concerned about the quality of their surface and ground water. What can it mean to advocate a market solution rather than regulation? Depending on your perception, some will assume that the neighbors must buy out the feedlot operators and others will assume that the feedlot operators must buy out the neighbors. But this is not a matter for assumption: it is a matter for collective choice and, ultimately, a moral choice of whom is to count.

By the time markets are operative, all of the tough moral choices structuring opportunities have been settled. Property rights are antecedent to markets (Schmid, 1999). Markets are not the place for deciding who has what to trade in the market. The reason anyone wants to be an owner–seller is so that they can coerce non-owners to pay them if the nonowner does not like the owner's use of the resource. If I am a nonowner neighbor, I don't really care whether the feedlot operator or the government tells me to keep off and not interfere. Ultimately, of course, it is both since I pay attention to owners and because I know that the government will back them up. Likewise, if I am a nonowner feedlot operator, I don't really care whether the neighbors or the government tell me to keep off the neighbors' surface or ground water where I seek to store my animal waste.

judgment, but it will not make the case for any particular judgment. I only want to note the argument of Callicott, who finds intrinsic values as a result of the proposition that "The existence of means . . . implies the existence of ends" (Callicott, 1999: 240). I observe that people do hold some values that they do not regard as instrumental to other values, and that they experience them as good in themselves. The problem is that different people hold different ideas of intrinsic values, which suggests that the value is inherent in people and not things. Our experiences are different. If you can persuade me that your experience is closer to nature, that may secure our peace. But if you fail, then our peace depends on the acceptance of a constitution in which we agree to play by the rules and will not burn the place down when we lose. It may be that self-limiting, loving people are the ones who have survived evolution to date, and we certainly have more bridges to cross as a result. But this is a kind of consequentialism as much as intrinsic value, and will not secure the peace among those who desire different consequences or who are experiencing different intrinsic values. Some philosophers argue against an anthropocentric ethic and ask, "Is not the ultimate philosophical task the discovery of a whole great ethic that knows the human place under the sun?" (Rolston, 1991: 96). Personally, I am not waiting for the truth to be revealed by economists, ecologists, *or* philosophers. We shall just have to work out our peace as best we can.

Cost and therefore economizing is not a natural phenomenon of the production function but, rather, an institutional artifact (Samuels and Schmid, 1997). What does it take to grow hogs? The recipe includes at least land, labor, buildings, feed, and medicines. Oh yes, it also includes a place to put the hog's waste. A place to put the hog waste is no different from a place to put the feed or a place to keep the hogs dry and comfortable. The only reason that the hog feeder pays for any of these ingredients is that these inputs are owned by someone else. The feeder would not pay for labor if people did not own their labor power. The feeder would not pay for buildings, feed, or medicine if they could be had for the taking. They are not for the taking if they are owned by others. This is true, as well, for surface and ground water. If they are owned by the feeder, then the feeder listens for bids from neighbors and compares the bids with the value of the resource in use. But if they are owned by the neighbor, it is the other way around. The ground water is no more (or less) an externality than the building. Incompatible uses are everywhere. Externalities, far from being a special case, are ubiquitous.

In all major and contentious environmental issues, an individual's willingness to pay, if a nonowner, and willingness to accept if an owner–seller, will be different because of wealth effects and transaction costs associated with the alternative assignments of ownership rights. Therefore, *economizing or maximization of total wealth cannot be a guide to the specification of property rights*. Specification of property rights affects the prices and values in any aggregated wealth calculation. Rights have to be worked out in a collective political process and ultimately are accepted (or rejected) on the basis of their ethical persuasiveness. So much for "free" markets. If they are markets and not war zones, they are never free for all. They are necessarily arenas in which some have freedom and some are exposed to that freedom.

Housing and landscape

The use of landscape for housing estates is greatly influenced by the happenstance of ownership boundaries. Most planners agree that clustering of houses is a desirable way to leave open space for wild lands and farming. These amenities make the house sites more valuable. Still, it is common to see houses on street grids that pay little attention to the landscape. Planners and home buyers may agree that a particular plot should be kept in wild land or farming, but if that is all a particular person owns, an agricultural zone is tantamount to saying that some owners get no land value appreciation from urbanization, while their neighbors with development permission become millionaires.

If the only way parcel owners can obtain land appreciation is when their particular parcel is developed, there will be great pressure on open space zones. The answer is strong regulation or transferable development rights. In effect, the appreciation potential is jointly owned by all landowners in a large area. And the owner whose land remains open still gets part of the gain when another parcel is built upon. Markets are still operative, but who has what to sell is different and the resulting landscape is different. It is a moral question whether we want to hold out to be the lucky owner with development permission and get all of the appreciation, or share the gain and create a different landscape and housing environment in the process.

Consumer ethics

Can ethical judgments be communicated in markets? Self-imposed industry standards, as an alternative to regulation, are now widely discussed. Industry (including that in the countryside) agrees to some standard of resource use that is more limiting than that given by their explicit property rights. The International Organization for Standardization (ISO) is one organization that facilitates self-regulation. However, at the present time, ISO 14001 only requires a management plan and compliance with national laws, but no specific environmental performance achievement. There is no link to any eco-labeling standard. Firms adopting the European Eco-Management and Audit Scheme (EMAS), which became effective in 1995, must publish environmental statements (Orts, 1995). The general idea is that firms adopting standards can then advertise the fact. Consumers of the products can then reward cooperators and punish others by withdrawing their patronage. We have seen Nike shoes change their labor practices in poor countries in response to consumer boycotts. We have also seen some furniture manufacturers advertise that they have obtained their wood products in a sustainable manner. Gerber, Frito-Lay, and Heinz advertise that they do not use genetically modified organisms in their food products.

Market-rewarded environmental practices hold some promise, but have limits. In this day when so much consumption is optional and image oriented, image is all-important, and the image can include concern for the environment. But, if consumers in the rich countries are not sympathetic to the interests of forest neighbors (or labor), it is hard for the latter to make much impression via the market. Coming back to the feedlot example, the neighbors do not want to depend on pork consumers to protect their ground water interests.

Ethics of uncertainty

Environmental issues are often quarrels over uncertain consequences of resource use. Returning to the feedlot example, one can imagine that the property rights to ground water were worked out in the state of North Carolina, which recently experienced rapid growth in large integrated feedlots. The state required certain waste lagoons that promised to protect the ground water. But some risk-averse people objected, because they feared that something might go wrong. They may not have been able to name what might go wrong, but they were just worried, and they would have preferred the environmental benefits of small operations to cheap pork. The unexpected was supplied by a hurricane in 1999. Its winds did not damage feedlots, but the torrential rain did. The floods killed many animals and flushed out many lagoons that carried this biochemical oxygen demand to the coastal water and estuaries. There have always been floods, but humans had not concentrated so many animals and so much waste in their path before. It is not clear what the ultimate result will be. Surely the risk-averse could have bought out the right of feedlot operators to create this possibility. Alternatively, the feedlot owners could buy out the risk-averse or assume liability for the damages when they occur. Of course, the firms then might be bankrupted and would not be able to pay catastrophic damages anyway. Who is free: the cheap pork lovers and producers, or the concerned environmentalists? This is a deep moral question.

Courts

In countries that follow the English common law tradition, the courts settle many conflicts over resource use. For example, a homeowner may object to the neighbor's animals, loud parties, satellite dish, storage of junk, or whatever. There may be no statutory law governing opportunities, but if there is a dispute, the courts will decide who is the right holder.[3] If there is conflict, there will be a property right (or there is war). The court will choose whose interests count by labeling the neighbor's use a nuisance or not. The neighbor is either free to proceed or the plaintiff is free of the offensive use.

The background provided by past rulings (and expected rulings) makes market exchange possible. You do not buy what you already own. If you like animals and loud parties and the courts regard them as a nuisance, then you are a buyer, and you must buy the right from your neighbor so that he or she will not sue when you proceed. On the contrary, if you like quiet, and noise is not ruled a nuisance, you must pay to keep your neighbor quiet.

Court-made law differs in the remedies that it offers for violation – either damages or injunction. Is there an economic basis for the choice of remedy that escapes the necessity of a moral choice? No. An opportunity backed by forcing the offender to pay damages is equivalent to private eminent domain. The offender may proceed and pay only the court-determined damages, which will be in reference to a generally observed market value. If the rights-holder has a unique valuation, it will not count. However, if the rights-holder has the remedy of an injunction, the person considering a grab must come to the rights-holder as a buyer and meet his or her unique reservation price. If you have a choice, you always want the remedy of an injunction in your opportunity portfolio. Whether unique preferences count is a moral question (Calabresi, 1985).

In many neighbor–neighbor conflicts, the parties are monopolists and no alternative suppliers are available. Mutually advantageous trade may not occur when personalities are involved. For example, an animal feedlot operator needs the local ground water as a waste sink and more distant sources are not substitutes.

---

[3] Countries such as France that follow the Civil Code rely more on the code and other statutes, rather than the *stare decisis* of case law. Still, courts must interpret and elaborate the law. "It is observable that entire bodies of law in civil law systems have been built up by judicial decisions in a manner clearly resembling the growth of Anglo-American common law. This is notably so, for example, in the case of French tort law" (Glendon, Gordon, and Carozze, 1999: 132).

If the feedlot operator is offended by an "unfair" monopoly price demanded by the ground water owner, the operator may abandon the project rather than buy the input.[4]

Tort law is a substitute for statutory law. There may be no zoning ordinance that prohibits animals and businesses in a residential area, but if there is a private dispute, the courts will decide who has the opportunity. It is common to hear people object to the volume of unnecessary statutory law. But it is a mistake to assume that if there is no statute, there is no law. The many volumes of case law are a substitute.

It may seem tedious to work through the details of formulating an administrative regulation. Would it be simpler to leave it to the courts? No. The same questions must be faced, and it is the detail that lawyers argue about when reasoning from case precedents. The court must form a standard for reasonable behavior – just as a legislature does. Liability is always defined against some standard of reasonable behavior. The standard may be what most people do on the average in a similar situation, or it might be the best available practice and technology. Definitions of negligence involve choosing sides, and choosing sides is a moral issue.

Can the government ever be absent when there is human interdependence and conflict? Can it keep out of it by remaining silent? No. Silence is no less a choice. If government fails to provide explicit rights, then the *de facto* use of a resource goes to whoever can physically take it. I call this "rights to the grabber." For example, if a farmer bulldozes a hedgerow and destroys a wildlife habitat, or fills a marsh with the same result, and if the court or legislature will not act, then the farmer has the *de facto* right and neighbors are exposed to the farmer's presumptive freedom. The farmer is in a much better position to grab than his neighbor. If the quarrel is about fish in a stream and the government is silent, then anyone with a fishing pole or net can grab. The destruction of the environment in a free-for-all is well known.

Even the *de facto* rights of the "grabber" are usually made effective by governmental limits on what aggrieved parties can do to avoid the grabber. Grabbing invites a counter-offensive. For example, if a wildlife lover is told that the farmer may employ any means to remove the hedges, then the wildlife lover may employ any means to prevent it. If this is just a state of nature, then the

---

[4] The following is an example of an agricultural nuisance case. In *Spur Industries v. Del E. Webb Development Co.* (1972) operators of a cattle feedlot were sued by a real estate developer whose house-building came closer and closer to the feedlot. The developer complained of odors and flies. The defendant argued that the developer had "come to the nuisance" and therefore was barred from relief. The Arizona Supreme Court granted an injunction conditioned upon the developer paying the costs of closing up or moving the operation. This means that feedlot operators do not have the right to set their own reservation price to sell out to those who object, but still have the right to pollute unless compensated at prices determined by the court. Is this rule more efficient than an outright injunction that would have required the feedlot to buy out the developer if it were to continue? Efficiency cannot be the guide, since law affects the prices. A feedlot can occupy a strategic location and make its neighbors pay its monopoly-like price rather than the going market price as seen by a court. Willingness to pay and willingness to sell are not equal.

bulldozer can be blown up. Such a state is one of Hobbesian war and no rights. The public must face its moral responsibility when it limits what one party may do to avoid the harm of another. This case is not idle speculation; witness Green organizations that employ violence.

What about implementing environmental control via private contract? If a developer owns a large parcel and sells lots to home-builders, the sales contract may provide that the land may only be used for certain purposes in the future. The city of Houston, Texas, has no public zoning, but private developers have encumbered the land because buyers like it that way – or at least some of them do (Seigen, 1970). However, the government is not absent from any contract. It has to decide whether to enforce different types of contracts and whether they continue indefinitely into the future. For example, a seller might contract with a buyer such that the buyer agrees never to resell to a member of a certain race. This contract would not be enforced in most jurisdictions. The participation of the public in contract enforcement is a moral issue.

Regulation

It is so easy to see government regulation as coercion. The regulated cannot do what they want to. But the other side is that if certain persons cannot do what they want, someone else is protected from their unwanted action. Alpha's nonopportunity is Beta's opportunity (Schmid, 1999). With scarcity and human interdependence, coercion is the inevitable accompaniment to opportunity and freedom.

Wetlands

Consider the example of wetlands. In the United States, filling of wetlands adjacent to navigable waters requires a permit from the US Army Corps of Engineers. Other wetlands are protected by local ordinances (if at all). Home site developers prevented from filling or dredging a wetland often protest that they have been deprived of the value of their property. But what constitutes *their* property is the very question at issue. For every wetland owner who finds the selling price of their land decreased by a regulation, there is another landowner who finds the selling price of their land enhanced by its access to ducks and fish, or is the beneficiary of water filtration. The beneficiaries may not literally be landowners, but their wealth and quality of life is nevertheless affected.

A regulation functions as a property right for the beneficiaries of the regulation. It is so easy to see the government as the beneficiary, which results in slogans such as "get the government off the backs of the people." But the people are not homogeneous in their preferences. More accurately, the only choice in an interdependent world is to choose which individual or group is on the back of another. If everyone can't have his or her first choice, someone will be on someone's back, and there will necessarily be coercion.

The role of selective perception in the framing of property rights issues was demonstrated in a survey of residents in a Michigan community that contained substantial wetlands (Pierre, 1999). When people were asked whether someone who destroys a community resource such as a wetland should compensate the community, three-fourths agreed that public compensation was appropriate. When people were asked whether the government should compensate a wetland owner if preservation is required, three-fourths agreed that private compensation was appropriate. These contradictory results illustrate how citizen choice of who is buyer and who is seller is influenced by who is most easily seen as the natural owner and what is perceived as the starting place from which change is initiated.

Fox hunting

Consider another countryside issue, that of fox hunting in the UK. This is an emotional issue for many countryside residents who have a social way of life built around fox hunting. It is equally emotional for animal rights enthusiasts. A ban on fox hunting would restrict the freedom of the hunter. Its allowance restricts the freedom of fox sympathizers. If the regulation had instead been an exchange right, whoever had the right could sell it to the other side if the price was right. But, as noted above, each side wants to be the seller. One characteristic of this and many environmental issues is that there are many people on each side. Thus, getting everyone's agreement to either buy or sell is costly. These transaction costs mean that whichever side is declared the owner is likely to remain the owner even if the sum of the willingness to pay of the buyers exceeds the summed reservation price of the sellers. Should the right be given to whichever side an economic analysis suggests would make the greatest offer gross of transaction costs? If that were the rule for determining property rights, the poor would be even poorer than they are today, since the rich would not have to pay them for what little they have.

There is one fundamental difference between a regulation that gives a use right and ownership in the form of an exchange right. Usually, the beneficiaries of a regulation – such as a ban on fox hunting or wetlands drainage – cannot sell the right even if the offer is greater than their reservation price. Note, however, that if the reservation price is greater than the offer price, the exchange or use right result is the same in terms of resource use. When there are many beneficiaries of a regulation such as a ban on fox hunting or drainage, there is usually someone who places a great value on the natural use of the resource. Because the good is nondivisible, those who place a low value can't sell independently of those who place a high value. A regulation then puts the persons with high value in the driver's seat. They would not sell even if they could. They are not missing out on any Pareto-better trade.

The Clean Air Act requires the US Environmental Protection Agency to set national ambient air quality standards. The Act has been interpreted to forbid the consideration of costs in setting the health-based standards. These provisions are referred to as "rights-based" because all individuals thereby have a right to a minimum level of protection (Powell, 1999: 10). The beneficiaries are not buyers but, rather, sellers, and the reservation price is infinite. The Food Quality Protection Act also forbids consideration of costs for chemical registration.

Transaction costs are often the functional equivalent of ownership. If Alpha and Beta do not trade because of transaction costs, Beta benefits if they prefer the present use of the resource. Transaction costs are not like friction, where everyone is in favor of grease. Some enjoy things being stuck in their present use. Thus, the ability to create transaction costs can be the means of implementing certain interests. Given the land-grabbing history of Ireland, the Irish constitution is very explicit about government taking private property. Any diminution of value as a result of land-use controls is interpreted as a taking. Is there then no effective public direction of land development? If the government would lose in a court challenge to denial of planning permission, how can any public purpose be implemented? The unofficial but functional answer is transaction costs. Appeals and court action take time. Time is money to a developer. In effect, the local planning authority bargains for conditions. If the developer is cooperative, the planning permission comes quickly. If not, it is fought to the bitter end, when the authority finally gives in rather than purchase the development rights. Transaction costs (called red tape if you oppose them) function as property rights.

There are many property rights equivalents. Regulation that gives options to some and exposures to others is a functional equivalent to rights of sellers–owners and buyers who are not owners in markets. Both require a moral judgment of whose interests count. Whether law is made by legislatures, administrative agencies, or courts, there is an inescapable moral question of who is a rights-holder. And the answer to this question drives what happens in markets by deciding who is buyer and who is seller of conflicting opportunities. Alpha can have an effective right, whether implemented by a legislative or administrative regulation or by a court decision on tort liability.

Public Spending, Taxes, Fines, and Subsidies

Government can affect the allocation and use of resources by the rules of markets or by directly producing goods and services. Take the case of birds that nest late in the spring in farmers' hay fields. There is an incompatibility between hay yield and bird yield. Late harvest increases nesting and number of birds reproduced and early harvest increases hay yield over the season. Depending on the rights (legislative or court-made), the bird lovers have to buy out the farmers or the farmers have to buy out the "birders." If the farmers have the rights, it is doubtful that the birders can collect the necessary purchase price from themselves, since the good has a high exclusion cost and invites free riding. If the free riders are not to defeat the project, the government must collect a tax to buy out the farmers. The method of finance itself raises moral issues to settle the conflict between those who would pay, but don't get the good because of equally benefited free riders, and those who don't want the good, but are forced to pay a tax.

Government spending to buy out the farmers could be evaluated to see if the benefits are greater than the costs. Benefit–cost analysis involves a whole host of other moral issues, especially if the distribution of income is questionable. This issue is addressed in other chapters in this volume (see also Schmid, 1989; Vatn and Bromley, 1994).

Is the payment to farmers a subsidy? It all depends on how you use the word. When a rights owner receives a payment in the market, we usually do not call it a subsidy. And it is certainly not a bribe. Whatever it is called, the payment is recognition of a right, whose taking must be compensated. On the other hand, if the government passes a regulation setting the hay cutting date such that farmers lose income, but offers no compensation, the government is saying that the public had an interest in this land (a kind of part owner) – and if any extra income was earned before, it was a gift from the public, which it no longer wishes to make. Who pays whom and who claims income is a matter of rights.

What about taxes? The early harvest of hay could be discouraged with a tax. It is easy to see taxes as just tribute to the sovereign and something apart from the people. But if bird lovers are rights-holders, then the tax has some similarities to a market payment. It differs in that the payment may not go to the bird lovers directly, although it could in the form of government wildlife management expenditures or improvements. It also differs in that the farmer gets to decide the use, which is again equivalent to private condemnation. The birders can't refuse the "sale" and the price is set by the level of the tax.

Pigovian taxes are often seen as a market failure correction. However, functionally, they don't correct markets, but determine who has what to sell in the market. A tax reflects an ownership claim. The word "tax" is actually a misnomer, for it is simply a payment for services rendered by the government.

The other side of a tax collected is a tax not collected. Governments are often urged to offer tax exemptions if a party will take a specified action such as conservation. This exemption is just the same as a payment and reflects an ownership claim. Such an exemption or subsidy is often billed as a voluntary alternative to regulation. From the taxpayers' perspective, there is nothing voluntary about it. They would prefer to be rights-holders and let the farmers in this case be the buyers, in the form of market payments or fines. Again, the question of who is a rights-holder is a moral issue.

Taxes, permits, and trading

The use of taxes to accomplish pollution objectives is much discussed but not widely used (although there are a few programs in Europe, such as Sweden's tax on SO_2). Regulations and permits – and more recently in the US, tradable allowances – are more common. The US has turned from contingent permits to permanent caps and trading for SO_2. The beneficiaries can be different.[5] A tax functions as a payment to a public resource owner, and the owner

---

[5] "Different instruments for pollution control have different implications . . . for the distribution of income within an economy" (Perman, Ma, and McGilvray, 1996: 233).

can subsequently raise the tax (reservation price) if it wants to lower pollution in the future or capture some of the windfall gains.[6] In contrast, a cap functions as a sharing of ownership, where the shares are fixed. The initial cap may be less than what polluters used in the past. But in any case, once the cap is set, it is fixed, and the only way the public breathers can get more is to buy it. This means that any gain in technology of production or treatment gives no benefit to the public, in contrast to the previous policy, which made the permit contingent on technology. Rights to benefit from technological change are a major factor in wealth distribution.

The cap and trade policy is much advertised for its cost saving compared to contingent permits (usually presumptively labeled "command and control"). A nontradable permit (a use right) results in resource use by its rights-holder even if other polluters put a higher value on it. The users with cheap options to using their permitted amount have no incentive or ability to sell their use to those with higher cost. Thus, the total cost of meeting the permitted level of pollution is higher than need be (Hanley, Shogren, and White, 1997: ch. 5).

Less attention has been paid to the distribution of the saving. It is distributed among industrial buyers and sellers. None goes to the breathers. Since the cap is fixed and the efficiency gains go to the polluters, it would not be surprising if the public who were suffering from the remaining pollution did not applaud the policy change. No matter how cheap the achievement of the cap becomes, there is no breather benefit (even if consumers gained, assuming pure competition). Breathers are not just interested in cost-effectiveness to achieve a given level of pollution, but mostly in environmental progress over time.

At first, environmentalists politically resisted the cap and trade policy because it appeared that firms were to be allowed to sell rights to pollute when environmentalists thought that they already owned the right. While the permitted level (prior to trading rights) could in principle have excluded new industry, in practice it did not do so fully. So, some environmentalists found the cap attractive in preventing new entrants from getting more of the public's share. And, if the cap was lowered beneath the historical level, the environmentalists gained in the short run. Perhaps the political compromise would have been different had the environmentalists seen that they were being asked to give away their opportunities for future improvement. If the parties do not understand the issue of how different policy details affect people differently, the political and ethical issues are not faced.

---

[6] An experimental program to phase down the use of chlorofluorocarbons and halons created a concern that the decreasing availability of ozone-depleting chemicals would raise their profitability to the rights-holders. The Congress imposed a tax on the remaining use of these chemicals to capture some of the windfalls. A tax approach has some advantages where new information and technologies are unfolding. "Marketable permits, where control agencies have the flexibility to vary the stock of licenses (as in the manner of open-market operations for short-term debt), or pollution tax instruments where tax rates can be altered as new information becomes available, offer attractive alternatives" (Perman, Ma, and McGilvray, 1996: 225).

Moral Choice Revealed

The above discussion illustrates that each policy instrument has a moral choice embedded in it. The point can be seen even more clearly if different instruments are applied to the same environmental interdependence (externality). Any instrument distributes rights among the interdependent parties and determines whose interests count. No instrument can avoid taking sides. Suppose that the moral choice is to favor environmentalists who want to avoid damage to surface and ground water by large-scale animal feeders.[7] The following demonstrates where the moral choice is embedded in each instrument expressed as ownership:

- *Markets.* The feeders are required to pay environmentalists for use of water for waste disposal. Feeders buy water disposal rights just like any other input to their production process that is owned by others, such as labor or machines. *The environmentalists are owners and possible sellers.* A so-called "cap" on animal waste would define that which is owned by environmentalists. If set at zero, environmentalists would own the entire resource.
- *Courts.* Feeders' use of water for waste disposal is declared a nuisance. If damages are the only remedy, environmentalists receive the market value as compensation, but have no right to refuse to sell. If an injunction is an available remedy, environmentalists may refuse to sell or will receive their own unique valuation even if greater than the usual market value. *Environmentalists are owners and possible sellers.*
- *Regulation.* Feeders are prohibited from discharging wastes into the waters. Environmentalists may either enjoy the use rights, as is usually the case, or they may be allowed to agree not to pursue their rights if paid (this depends on whether the public authority can act on its own or only on initiation of a complaint by the beneficiaries). *Environmentalists are owners and possible sellers.*
- *Public spending and taxation.* The feeders might be taxed to discourage use. While it is often called a tax, it functions like any market fee for service. Even if the environmentalists do not receive the revenues, *they are owners* to the extent that feeder use is reduced. Which environmentalists count depends on how the tax level is set. Some will favor a tax that is high enough to discourage any "sale" to and use by feeders. Alternatively, if feeders own and the environmentalists want less pollution, they have to buy out the feeders. This might take the form of "subsidizing" the feeders to contain their waste, this being financed by an income tax. This is functionally equivalent to a market payment to the feeders that they may accept or refuse.

In practice, these policy instruments/institutions are complements as much as substitutes. It is legislative regulation and court decisions that are the foundation

---

[7]   The claimed substitutability among policy instruments is in terms of consequences. Philosophers distinguish consequentialist (utilitarian) from rights-based policies (Thompson, Matthews, and van Ravenswaay, 1994: 82–103). A particular expression of a right may be prized in itself independent of its consequences, in which case different institutions are not substitutes for some people.

of the market, as well as public spending. What the government pays for as the agent of an interest group depends on who the rights-holders are. And the question of who the rights-holders are is an ethical issue. To say that policy instruments are substitutes is not to say they are perfect substitutes. One form or the other can make the preferences of particular parties stronger than an alternative.

Institutional Change

While different policy instruments can produce similar results, they are not equally available to different interest groups. Some groups are more powerful in their influence on legislatures, some at the local level, and some at the national or international level (Heretier, Knill, and Mingers, 1996). Other groups have a greater ability to initiate court action or administrative rulings. And some groups are in a better position to effect institutional and legal change by affecting the evolution and application of ideologies and unavoidable selective perception. All of this depends on the political rules for making ownership rules. The choice of political rules that influence the ability of groups to get their preferences to count is another arena of ethical choice.

Institutional change can be illustrated with the case of animal feedlots in the US. Until 1977, the only remedy for those damaged by this kind of pollution was in private court suits under the common law of nuisance. The speed and path of court-made law was not acceptable to some interests and so they turned to Congress. The Clean Water Act of 1977 defined animal feedlots as point sources that are similar to industrial outfalls that require permits. Administration was left to the states, and no permits were issued until the early 1990s. Even today, many states have no permitting program for feedlots. One can say that either the moral judgments or the political power differ among the states.

Meanwhile, powerful large-scale animal feeding interests have passed right-to-farm laws that exempt agriculture from nuisance law in many states. But, in Iowa, a small-scale farmer sued the government when a neighbor was about to build a large smelly operation next to his farmhouse. Namely, in *Bormann v. Board of Supervisors* (1998) the Iowa Supreme Court declared the state's right-to-farm law an unconstitutional taking of the right to be free of nuisances. Different interest groups can change or block rights changes at different levels.

Conclusion

The argument has been made that the choice of environmental policy instruments unavoidably contains a moral judgment. *There is no tradeoff between economic efficiency and moral principles.* Economic efficiency is not a single entity. There are as many efficient outcomes as there are property rights distributions that implement some moral choice. Change the distribution and you change what becomes the efficient outcome: "Efficiency criteria are determined relative to an existing system of rights and privileges. They provide no basis for evaluating proposals to change that system" (Thompson, 1995: 109; see also Randall, 1972).

If moral choices must be put front and center, then what moral principles are to be embraced? I have my preferences, but I claim no special status for them as an economist. I only want to participate with others in some political working out of competing claims and their moral base. Economists and others can clarify the issues, as I hope I have done. These moral issues and choices are before us and have not already been made: they only await clever discovery. I'm sorry, but we are all going to have to work at it. This is not a matter of reading the tea leaves of society or looking to science for the answer (Busch, 2000). It is a matter of choosing who is my brother and sister – who is a subject worthy of standing in the sun and not an object to be manipulated. It is ultimately a matter of choosing how you will limit yourself in your transactions with others. A claim of a right with no willingness to self-limit is hopeless, since no one can grant it without a similar self-limit. Persons who are unwilling to limit themselves can hardly ask others to do it.

The argument of this chapter has two primary implications for policy-makers. Since policy-makers are also teachers and part of the learning environment for their constituents, they have a role to play in clarifying value conflicts. Policy-makers can help people to see that government has a face and is not an abstract irritating force. *The face of government is the face of one's neighbors.* Seen in this way, the necessity of moral choice is clear. The choice is not government versus no government, but whose side government must inevitably take. Policy-makers can also help people to see that to demand rights requires self-limits. In closing, I quote from the philosopher John F. A. Taylor:

> I am as aware as another that the political condition of mankind . . . is discovered always in a context of struggle. . . . You shall not legislate in a free society whether there shall be struggle. You can legislate only that the struggle which occurs shall be human. A struggle is a human struggle not because power is loosed in it, but because persons are party to it. . . . In the exercise of that freedom lies whatever is human in ourselves, and if you are diffident of that freedom, that is a delusion of the ignorant or a sentiment of the forum, then, whatever it is you discuss, you do not discuss the historical condition of men. Which is simply to have said, you do not discuss politics. You may reject the tragic risk which another has imposed. . . . But if you reject the risk which you yourself impose, you do not secure your humanity; you desert it. (1966: 294–96)

References

Bormann v. Board of Supervisors, 584 N.W. 2d 309, 1998.

Busch, L. 2000: *The Eclipse of Morality: Science, State and Market.* Hawthorne, NY: Aldine de Gruyter.

Calabresi, G. 1985: *Ideals, Beliefs, Attitudes, and the Law: Private Law Perspectives on a Public Law Problem.* Syracuse, NY: Syracuse University Press.

Callicott, J. B. 1999: *Beyond the Land Ethic: More Essays in Environmental Philosophy*. Albany, NY: State University of New York Press.

Glendon, M. A., Gordon, M. W., and Carozze, P. G. 1999: *Comparative Legal Traditions in a Nutshell*. St. Paul, MN: West Group.

Hanley, N., Shogren, J. F., and White, B. 1997: *Environmental Economics In Theory and Practice*. Oxford: Oxford University Press.

Héritier, A., Knill, C., and Mingers, S. 1996: *Ringing the Changes in Europe: Regulatory Competition and the Transformation of the State: Britain, France, Germany*. Berlin: Walter de Gruyter.

Orts, E. W. 1995: Reflexive environmental law. *Northwestern University Law Review*, 898(4), 1227–340.

Perman, R., Ma, Y., and McGilvray, J. 1996: *Natural Resource and Environmental Economics*. London: Longman.

Pierre, K. 1999: *The Susceptibility of Property Rights Heuristics to Framing in Public Opinion Polls and Voting*. See http://www.msu.edu/user/schmid/pierre.htm

Powell, M. R. 1999: *Science at EPA*. Washington, DC: Resources for the Future.

Randall, A. 1972: Welfare, efficiency and the distribution of rights. In G. Wunderlich and W. L. Gibson, Jr. (eds.), *Perspectives of Property*. University Park, PA: Pennsylvania State University Press, 25–31.

Rolston, H. 1991: Environmental ethics: values in and duties to the natural world. In F. H. Bormann and S. R. Kellert (eds.), *Ecology, Economics, Ethics*. New Haven, CN: Yale University Press, 73–96.

Samuels, W. J. and Schmid, A. A. 1997: The concept of cost in economics. In W. J. Samuels, S. G. Medema, and A. A. Schmid (eds.), *The Economy as a Process of Valuation*. Cheltenham, UK: Edward Elgar, 208–91.

Schmid, A. A. 1989: *Benefit–Cost Analysis*. Boulder, CO: Westview.

Schmid, A. A. 1999: Government, property, markets . . . in that order . . . not government versus markets. In N. Mercuro and W. Samuels (eds.), *The Fundamental Interrelationships Between Government and Property*. Stanford, CN: JAI Press, 233–7.

Seigen, B. H. 1970: Non-zoning in Houston. *Journal of Law and Economics*, 13, 71–148.

Spur Industries v. Del E. Webb Development Co., 494 P.2d 700 (Ariz. 1972).

Taylor, J. F. A. 1966: *The Masks of Society*. New York: Appleton–Century–Crofts.

Thompson, P. B. 1995: *The Spirit of the Soil: Agriculture and Environmental Ethics*. London: Routledge.

Thompson, P. B., Matthews, R. J., and van Ravenswaay, E. O. 1994: *Ethics, Public Policy, and Agriculture*. New York: Macmillan.

Vatn, A. and Bromley, D. 1994: Choices without prices without apologies. *Journal of Environmental Economics and Management*, 26, 129–48.

CHAPTER **10**

Efficient or Fair: Ethical Paradoxes in Environmental Policy

Arild Vatn

The environment creates special challenges for economic theory. It has characteristics that do not fit well with the standard presumptions of the neoclassical model. Goods are complex and characterized by physical interdependencies. The interactions between the economy and the environment are also characterized by fundamental uncertainty, novelties, and large time lags. Finally, effects of actions undertaken at one place are dispersed through the environment to other parts of the economy, often far away from where they originated.

All these phenomena have the potential of creating inconsistencies in an economic model based on independence and thus easily demarcated objects. The aim of this chapter is to analyze a specific set of such inconsistencies – those that challenge the distinction between efficiency and fairness, which are so important to modern welfare economics. These inconsistencies emerge at all three levels of economic analysis central to making environmental policy. First, they appear when we attempt to define "the optimal amount of nature." Second, they surface when we try to specify responsibility and draw distinctions between "who has the right" and "what is efficient." Finally, the characteristics of environmental goods create ethical problems when we search for the "least cost abatement strategies."

In conventional economic theory, the Pareto principle is used to establish a demarcation line between efficiency and distributional issues. Its introduction was an important step in the process of trying to make economics a value neutral discipline – to escape the problem of making interpersonal comparisons. However, the Pareto principle demands socially accepted rights as a basis, and there are disputes within economics about how far the neutrality of the conclusions reaches (Mishan, 1980; Schmid, 1987; Bromley, 1989; Calabresi, 1991; Griffin, 1995). The Paretian logic defends the status quo. Thus, one may claim that the Paretian rule – instead of guarding against value judgments – is, rather, disguising ethical issues.

This chapter accepts that there are considerable ethical problems with the Paretian optimality rule even when it is applied to ordinary market transactions

(Schmid, 1987; Bromley, 1990). However, this chapter focuses on the special problems related to efficiency and fairness in the field of the environment. Transferring the *commodity* vision from the marketplace to the environment substantially exacerbates some ethical problems inherent in economic theory.

To create a basis for the analysis, I will first look at the relationships between the concepts of efficiency and fairness in general. Thereafter I will examine the three levels of analysis relevant for environmental policy as briefly outlined above. In the conclusions I will bring together the observations made when examining the problem at these three levels.

Efficiency and Fairness

According to Elster (1992), principles of *fairness* or justice are applied when allocating scarce resources, necessary burdens, rights, and duties to different individuals in a society. At the same time, the allocation of scarce resources is the core concern for the theories of economic *efficiency*. It is no wonder that the issue of fairness and efficiency – their interdependencies and conflicting interfaces – has a long history in Western philosophy in general, and in economic theory in particular.

The modern discussion of fairness or justice originates from the writings of Hobbes, Hume, and Kant, and is continued today by, for example, Rawls. Hobbes and Hume advanced the view according to which justice is secured if rules give *advantage to all parties involved*. According to Barry, "justice is the name we give to the constraints on themselves that rational self-interested people would agree to as the minimum price that has to be paid in order to obtain the co-operation of others" (1989: 7). Kant and Rawls have suggested the principle of impartiality – that justice applies to *rules beyond the self-interest* of individuals. Thus this second approach "is not constrained by the requirement that everyone must find it to his advantage to be just" (ibid.).

However, it is difficult to draw a clear line between individual advantage and impartiality. As Barry (1989) explains, there are elements in the writings of Hume that relate to the idea of impartiality. Furthermore, the Rawlsian "veil of ignorance" is a way to combine the two perspectives. That is, behind a "veil of ignorance" individual advantage is transformed into impartiality. It has become a question about which rules would be universally accepted when no one knows which specific interests they would come to hold in the society, since they do not know which position or capabilities they will actually be allotted.

One may further divide theories of fairness or justice into consequentialist/welfarist, rights-based, and procedural theories. However, sharp distinctions are difficult to make. Welfarist theories focus on individual advantage, but cover large subgroups. A theory is welfarist if the only consideration relevant for the allocation of goods is their effect on the individuals' welfare or utility. Classical and ordinal utility theories are the main welfarist theories of fairness and justice. Yet they represent quite different positions, since the first accepts interpersonal comparisons while the second does not (Sen, 1979; Elster, 1992).

A specific issue of importance when evaluating consequentialist theories is the role that one grants to motives. Consequences may differ from what was intended. When formulating institutions, one may want to differentiate between intended consequences – those that could or should have been foreseen – and those that could not be expected. Should rules of action or reaction depend on the motivation behind the choices made? This issue becomes significant especially when we look at environmental problems, which often are viewed as unintended.[1]

Rights-based and procedural theories may either focus on individual advantage or they may be characterized by impartiality. Focusing on rights is often understood as an opposite pole to that of consequentialism. From a deontological viewpoint, such an understanding is warranted – a right is not to be violated no matter what the consequences. However, emphasis on rights should not be construed to always trump consequences.[2] Sen (1987) discusses this from the opposite angle. In his plea for rights-based consequentialism, he emphasizes that there need not be a conflict between focus on consequences and on rights. This is the case if one accepts to measure consequences along other dimensions than just welfare, such as rights distribution.

Procedural theories focus on the process chosen for determining outcomes. Taken to its extreme, the theory implies that outcomes are of no significance for the building of institutions. This is pure procedural justice. However, we also have perfect and imperfect procedural justice (Rawls, 1971), according to which procedures are formulated in order to obtain certain outcomes with certainty or with an accepted level of probability. One group of procedural theories emphasizes the right of *participation* as a goal in itself. This is especially accentuated by some as important in the formulation of policies pertaining to environmental risks. One prominent example is Beck (1992) and his study of the "risk society."

Principles of fairness may be characterized by high degrees of universality, as in Rawls. They may, however, also be "*local*," implying that they are related to specific cultures, sectors, or communities. Both Elster (1992) and Walzer (1983) argue that justice or fairness is normally of the local kind. Situations vary, and our perceptions, values, and norms are context-relative and dependent on historical conditions.

In the "local" sphere of economics, the most important rule of fairness or justice is the Pareto principle. It appears in different forms – as Pareto optimality, Pareto improvement, and potential Pareto improvement.[3] All of these measures are based on the criterion of individual advantage. Furthermore, they all presume rights to be previously defined. They are welfarist and considered universal.

[1] Economics regards externalities as unintended by-products of economic activities (Mishan, 1971; Baumol and Oates, 1988). For a critique of this view, see Vatn and Bromley (1997).
[2] A well-known example is the rule of not killing others (a right to life). What if killing one person could save ten other lives, or 100, or 100,000?
[3] Pareto optimality means that nobody's utility can be increased without decreasing the utility of others. A Pareto improvement implies that the utility of some agents can be increased without reducing the utility level of others. Finally, potential Pareto improvement means that some can gain while others lose, but that the gainers could compensate the losers and still be better off than before. It is not assumed that compensation is actually made.

Social welfare is solely based on (the sum of) individual welfares. When trying to establish a demarcation line between efficiency and fairness, all of these rules have to take the distribution of resource endowments for given.

While the principle of "advantage for all" may seem innocuous and simple, the Pareto principles have created a lot of debate and interpretations. Thus Griffin notes that Pareto optimality "should not be confused with the Pareto improvement criterion. Pareto improvement does not employ arbitrary utility levels for all individuals; it uses current utility levels" (1995: 4). Certainly, the whole grand utility frontier is Pareto optimal. However, this has no relevance in particular policy situations, with their rights and endowments structures. Thus both the Pareto optimality and the Pareto improvement rule have to reflect status quo rights. As is emphasized by Bromley (1989), even the form and position of the utility frontier is dependent on the status quo rights configuration. When applying Pareto optimality as a criterion, distribution has to be defined either as a noneconomic problem or circumvented by presuming the distribution to be optimal at the outset.

While Mishan (1980) is also supportive of the resemblance between the logic of Pareto optimality and Pareto improvement, I find it difficult to embrace his claim that "there can be little doubt that the adoption of an *actual* Pareto improvement, one that makes 'everyone' actually better off than he was before, as the norm of economic efficiency would be ethically acceptable to society" (Mishan, 1980: 148). Note that Mishan reformulates the Pareto improvement rule by presuming "everyone" to gain. Still, it is only by assumption that one can make even such an interpretation ethically neutral or indisputable.[4]

Moving to the potential Pareto improvement rule, it introduces further problems because it abandons the "advantage for all" basis. The introduction of cardinality (monetizing) and additivity causes additional problems. For Mishan, it is the move from Pareto improvement to potential Pareto improvement that represents ethical difficulties. Along the same line of reasoning, Griffin emphasizes that exchanging "harm for some people in return for 'greater' help for others" (1995: 14) is ethically problematic. He also shows that the results of a potential Pareto improvement calculation are strongly dependent on which utility levels – the initial or final, or any one in between – are assumed as the baseline for the comparison.

Thus we have to accept that the concept of economic efficiency is normative in a double sense. First, it has to accept some initial distribution as a base line. Second, it will prefer the action that creates the greatest net value – given the baseline – whatever the distribution of potential gains and costs may be. The alternative to the unsuccessful separation of efficiency and equity is to explicitly

---

[4]   Calabresi (1991) understands the Pareto test as a unanimity rule. From this viewpoint, Mishan is right, since no change will take place until everyone accepts it. All accepted choices pass the Pareto test by definition, whatever basis the individuals had for their acceptance. However, if we accept that the concept has a content – an advantage "for all" – we must also accept that some people may support a rule that does not promise them such an advantage. The maximin rule proposed by Rawls is such a rule.

open up for elaborating on the issue of which interests should be favored, and thus take a stand in relation to which direction economic development should take.

In our case, the most important issue is the effect the chosen rule or model has on what become relevant policy problems. The individualistic, universal, and welfarist viewpoint allows us to observe a certain set of problems. The opposite standpoint – that fairness has to be based on socially accepted interests (impartiality) and on procedures – draws the attention in a different direction. The relevance or consistency of these perspectives depends not only on the ethics that underpins them. It also depends on the character of the problem to which they are applied. As we shall see, rights and procedures become core issues in the case of environmental problems. To try to deny or circumvent this only creates confusion.

The "Optimal Amount of Nature"

As has already been emphasized, the physical environment has characteristics that deviate widely from the conventional assumptions about goods in economics. The environment is a complex system of interacting processes – vast cycles of matter and energy occurring over various scales of time and space as the result of myriads of self-organising processes (Graves and Reavy, 1996). Human intervention alters these flows either by accelerating existing ones, as in the case of nitrates and carbon dioxide, or through various transformations, finally emitting matter that is unfamiliar and poisonous to the system, as in the case of dioxins and PCB.

Due to the dynamics of a complex structure like an ecosystem, its changes and degradations are "novelties" rather than "negative externalities." The reactions of such a system to changes in its state variables will be difficult to foresee, and may often be recognized long *after* the actions that have caused the problem were taken (Perrings, 1987). An activity not thought to influence the possibility of undertaking other actions in the future may, after all, constrain future possibilities. Time lags are important, because they translate into lack of knowledge about future events related to complex systems. They may also have important effects on what becomes the "optimal amount of nature."

In what follows, I will analyze ethical paradoxes related to time lags in three steps. First, I will focus on the effects of investments in commodity markets on what becomes an "optimal amount of nature." Second, I will expand the argument to a more general thesis about incentive asymmetries between the market and the environment. Finally, I will discuss the consequences of physical interdependencies for the independence of individual preferences – the values that form the basis for decisions about "optimal amounts of nature."

Ex post and *ex ante* efficiency of market investments

The "optimal amount of nature" – the "optimal amount of pollution," for example – is defined as the point at which marginal environmental costs equal marginal costs of maintaining environmental qualities. The latter may consist of abatement costs or the costs of a reduced level of production. As market economies develop their forces of production, investments will be made in the production of goods to be sold in the marketplace. Since production directly influences the quality of the environment, these investments also influence what, at each point in time, become the costs of maintaining environmental quality.

There is thus an important asymmetry here, the consequences of which depend strongly on the prevailing rights structure, time lags, and knowledge. Let us start with the potential Pareto improvement as the efficiency rule. At a certain point in time, the environment is observed to degrade. Later, the degradation is linked to emissions from a certain industry, which has developed to rather comprehensive proportions over time. Economists are now asked to calculate the optimal level of pollution. Assuming that it is possible to estimate the environmental effects in monetary terms with the necessary precision, they soon find out that the costs of abatement exceed all involved environmental values. The value of the already undertaken investments – made on the presumption that no harm would be done – is very high due to high expected returns from future sales. Thus, the conclusion is that to do nothing is optimal. The emissions are deemed to be Pareto-irrelevant. Even if the case is not this extreme, existing investments in production will drive the conclusions when decisions on what to do can be made.

Could a retrospective Polluter Pays Principle (PPP) break this dynamic? This is not straightforwardly clear. One way to operationalize this rule would be to claim that the value of existing investments in a polluting industry should not be taken into account when defining what it is optimal to do. Put differently, if one already at "day one" had known the environmental costs that would appear at later stages, it would have been possible to correct the investments so as to make them "optimal" from the outset. Should this counterfactual situation be the basis for the procedures that operationalize the PPP?[5] If we adopted the retrospective PPP rule, one could ask on what basis would it be possible to determine the alternative developments in investments and technology.

The foregoing analysis demonstrates how optimality is based on an exceptionally fragile structure of assumptions. It also shows how what becomes efficient depends crucially on how we operationalize the PPP rule. Thus it is both very important and difficult to handle the structural asymmetry between *ex ante* and *ex post* efficiency. The arguments concerning the costs of doing something about the causes of global warming illustrate this. Normally, the reduced value of existing investments in production and infrastructure of taking action is calculated on the basis of prices determined before the new policy is set into motion – on the basis of the "old" rights structure. This gives "full power" to the past. When considering future costs, net present value calculations are invoked, with

---

[5] This may also provoke a debate about whether the PPP means that the emitter should pay compensation. Following the arguments of Pezzey (1988), the economist would find him- or herself in the following situation: Does the PPP imply that polluters should only pay the (optimal) abatement costs, or should they also pay for the environmental degradation? The definition has to be made outside economics, and that definition determines what becomes "efficient."

their effects on the balancing of future costs and gains. Another asymmetry related to time is present: abatement costs (costs of redirecting production and consumption) are incurred early, while environmental gains are obtained later. There is no wonder that "optimal" CO_2 reductions are normally calculated to be low (see Nordhaus, 1993). To the degree that such analyses involve discounting of *utility* across generations, they are counter even to the utilitarian ethics of maximizing the sum of utilities. A rule based on discounting the utility of future persons is illegitimate within most known ethical systems (Mishan, 1975; Spash, 1993), although many economists do not seem to be aware of this.

The argument that the future will gain from investments in man-made capital is invoked against arguments that criticize discounting. Future income will grow at least at the speed of the chosen rate of discounting. This may be a legitimate argument for discounting income. Still, even this hinges on brittle postulates. Here, we enter the debate about weak and strong sustainability, and the problems of substituting man-made for natural capital (Perrings, 1997; Nöel and O'Connor, 1998). The main problem in this case is how the apparent certainty of values flowing from existing markets will dominate over the uncertainty about what may become lost environmental opportunities in the future. Given the complex dynamics of systems such as the atmosphere, one can say very little about the future on the basis of evaluating the present and its history.

The asymmetry of incentives

The arguments discussed above may be expanded to a more general thesis about how incentive asymmetries influence what becomes "efficient." Often, our policy problem is to find a reasonable balance between market expansion and the protection of the environment. However, the two arenas are quite different in terms of the incentives they give for individuals. Environmental protection requires collectively agreed-upon restrictions on individual behavior, while market expansion is driven by the promise of economic gain for individual agents. That is, environmental protection is defensive and emerges a long time after problems are caused, whereas market expansion is inventive and is often the cause of later needs for environmental protection. As long as agents are individualistically rational, creativity will be attracted to the potentials of the market sphere. Thus, it is not only time lags that protect whatever investments have been made: investments will in general be directed to expanding the market.

The bias toward market expansion may sometimes be neutral to the environment. Still, as long as development is based on the use of resources, it will translate into detrimental environmental effects through, for example, changes in flows or matter or use of space. Following Perrings (1997), interventions into an ecological system will reduce its resilience and cause costs in the future. Thus, expanding markets can only by rare coincidences foster co-evolution – that is, reinforce natural dynamics. The argument is illustrated by the case of genetically modified organisms. The current expansion of patenting evidences a strong drive toward developing property rights and market institutions to create incentives for technological change. This is considered necessary because an important feature of biological material is its capacity to copy itself. However, this capacity also makes genetically modified organisms threatening to environmental systems: it makes the genetically modified organisms capable of spreading to natural systems and inducing changes in them.

Again, the incentive asymmetry will influence what becomes "an optimum." From the viewpoint of market incentives, motivational neutrality can only be obtained by making all natural goods marketable. However, this is not possible. As long as demarcation is not complete, problems will appear later elsewhere in the complex natural systems (Vatn, 2000). The real ethical problem is to define institutions that can balance *ex ante* the expansion of markets and environmental protection. This cannot be done on the basis of an efficiency calculus. There is – as we have seen – no fixed point from where to make that exercise, and it will end in circular reasoning. As far as I can see, this implies that society has to define some environmental standards that it devotes itself to.

This echoes the arguments for a Safe Minimum Standard (Ciriacy-Wantrup, 1968), which seeks to secure resource allocations that cannot be justified on efficiency grounds. However, my argument goes further, because it does not only attach weight to choices that may cause great and irreversible environmental damage in the future (see Norton, 1992; Toman, 1994). I argue for deliberately *choosing a development path* for the economy on the basis of *ex ante* definition of future rights and needs. This certainly takes us beyond the level of individual or self-oriented preferences.

Physical dependencies and interrelated preferences

There are two issues concerning preferences that are of importance for our case. The first relates to the problem of basing evaluations only on individual preferences. The second has to do with the ethical consequences of the interrelationships between the preferences of an individual and the opportunity sets that they create for others when fulfilled. My argument is that the choice of a developmental path for the society also influences the development of preferences. The culture or type of society influences desires of individuals living in it (Norton, Costanza, and Bishop, 1998). A related idea is that we learn what are attainable preferences (Elster, 1983). The development path will influence who we are becoming (Parfit, 1983). Choosing the direction for the economy can thus not be based on preferences that we have as individuals at an arbitrary point in time. These kinds of choices can only be made as the result of a continuous public dialog and reasoning.

Neoclassical theory does not recognize this problem. Either it postulates preferences to be given and stable (see Becker, 1976) or preference formation is defined as the business of other disciplines such as sociology. The latter view is inconsistent to the extent that it, at least implicitly, implies an acceptance of the social character of preferences. The only position consistent with the individualist perspective of standard welfare theory would be to claim preferences to be nonsocial. Otherwise, the question about which preferences are the best for us to hold immediately enters the stage.

The importance of the interrelationships between preferences, culture, and economic development becomes apparent when we appreciate the nature of environmental regulations. The environment is a common good in a specific sense of the word. My choices and actions related to the environment influence you whether you live today or generations later. If I prefer environmental goods, I will make your life better if you also prefer them. My willingness to devote resources to environmental protection will add to the aggregate willingness to do so in the society, which in the end determines the magnitude of the good available for all of us.

Is it irrelevant to you what preferences I have when what I value influences the costs that you face when trying to reach your goals? Certainly not. Should I take your interests into account when I form my preferences and consider my choices? These questions are also relevant when we consider "ordinary commodities." Given that there is scarcity, one person's right influences the opportunity set of others, because the right imposes a duty on them (Schmid, 1987). Physical interdependencies with respect to the environment add important new dimensions to the problem. For example, in the case of common pool resources (Ostrom, 1990) we "consume" the same good. What I value will influence your opportunities directly through my influence on the size and status of the good. Physical interdependence creates a situation in which everyone must ask himself or herself how and to what extent he or she *should* participate in the realization of the common good. This implies both defining the good and realizing what is defined.

According to Norton, Costanza, and Bishop (1998), understanding preferences as social is good news: "It may be possible to make a small social investment that will affect which types of consumption bring enjoyment to consumers, reducing the scale of human impacts without decreasing, perhaps even increasing, levels of welfare of consumers" (1998: 203). This view turns the attention toward communication about which preferences to hold. Economists tend to view this as not democratic (see Randall, 1988). However, if preferences are social, defending democracy becomes an issue of which procedures guarantee a democratic communicative process. Therefore, it could be considered an obligation to discuss which possibilities, norms, and preferences to hand over to future generations – to cultivate preferences in future generations (Norton, this volume). The interdependence of our choices and the opportunities of future people leave us no other choice.

Preference formation is a genuine social issue, which cannot be settled on the basis of individual willingness to pay. For example, prices cannot determine what is to be priced. When the good has to be collectively defined, as in the case of the environment, individuals are linked together through the decision-making processes and collectively agreed-upon decisions. This cannot but influence also the processes of realizing the common good.

In a situation characterized by physical interdependencies between individuals, a set of ethical considerations are "forced upon" each of us. It is in the interest of each individual that these issues are settled collectively. We can present our own views, needs, and desires to influence the collective process, and we also have to accept the interest of others to do the same. The ethics involved in this situation is different from pure altruism: it is best characterized by the concept of solidarity – supporting others under the assumption that it also benefits oneself, while recognizing that it may change one's own interests in the light of the needs of others. The practical policy problem is to develop institutional structures that facilitate solidarity and adjustments toward mutual interests. In my understanding this is the reason why procedural dimensions seem so central in the analyses of environmental policies conducted by noneconomists.

Whom is it Efficient to Blame?

The characteristics of environmental problems – especially the time lags – also cause difficulties when we try to draw a distinction between efficiency and who is or should be made responsible for any harm done. We shall discuss this both in relation to the Paretian principles and the issue of efficient incentives.

Along the path of development new resources will continuously come into existence and new emissions and environmental problems will occur. Therefore, there is demand for continuous formulation and reformulation of rights as well. This vests the polluters with *de facto* liberty, because rights and duties usually have to be determined *ex post*. Although environmental economics is based on the idea according to which emission taxes or tradable quotas can sustain Pareto-optimal conditions (Baumol and Oates, 1988), the Pareto improvement rule has little to offer here. Given the existing allocations at the time when an externality is observed, someone will lose. Only the potential Pareto improvement rule can thus be employed, with its problems related to value neutrality and ethical commitments as previously discussed.

While the retrospective Polluter Pays Principle (PPP) did not help us in deciding what is an "optimal amount of nature," it could still be thought to produce the right incentives here. This rule would make the polluter responsible for all emissions that later may be observed to cause environmental stress. Given this rights structure with its attached procedures, an "advantage for all" type baseline may be considered to be reinstalled. Again the hope is unjustified.

First, retrospective PPP will generate disincentives – it induces firms to choose short lives in order to relieve themselves from potential burdens. The rule thus tends to undermine itself. Second, there is the issue of acting in "good faith." This is not a trivial point if environmental problems are true novelties riddled with uncertainties or lack of knowledge. Emissions of carbon dioxide from fossil fuels have, as an example, a history of over 250 years. Still, we have not yet reached a conclusive understanding on their damaging effects.

Transaction costs – especially high costs of acquiring information – may also favor the status quo under retrospective PPP. This informational asymmetry may result in a moral hazard. Firms may choose not to conduct any *ex ante* evaluations of problems that their activities may cause in the future, in the hope of getting relief on the basis of good faith if adverse consequences surface later. However, how can we distinguish this case from cases in which the agent had no reasons to believe that problems will occur? This boils down to a question of whether – even in retrospect – only consequences should count, independent of motives.

We have to conclude that the PPP is not an efficiency rule at all, but a rule about whom is to blame. It is a rule about fairness, which may not always result in the least-cost abatement strategies. This relates to victims' defense as discussed by Coase (1960). Coase showed that, given positive transaction costs, it would be economically optimal for society for the polluter to do nothing when defensive activities are cheaper than abatement. The case of "the moving victim" reciprocity is emphasized. The identification of an emission and an emitter is not enough to specify an externality. For an externality to exist, both parties have to be present. Thus, responsibility cannot be determined only on the basis of physical relationships, as is implicit in the PPP. However, it cannot be determined on cost grounds either, as seems to be the Coasean position (Vatn and Bromley, 1997). Specific – local – evaluations based on relevant ethical considerations have to be undertaken.

Given the character of the problem, we have to ask whether responsibility can only be considered individual. Who is the emitter, after all? Technological change is predominantly a social process. By this, I mean that technological development is based on public investments and promoted through public education and extension. This is partly because of the public goods aspect of technology and to reduce transaction costs of implementing solutions that are considered favorable. If the society has supported a certain developmental path, it may be considered false or unfair to characterize all negative effects as results of individual decisions only. One could argue with equal strength that the observable problems are engendered by institutional arrangements, instead of being brought about by the intentional action of the involved individuals. Thus, if certain institutional arrangements are favored to overcome problems of applying the market model in the real world – a world in which interdependencies are ubiquitous and transaction costs are positive – it does not seem consistent to return to the market model when evaluating problems and ethical dilemmas produced by real structures that, in the face of things, have to deviate from pure atomism.

In the end, the ethical implications of the PPP may vary substantially across cases. This presents the question as to whether issues such as motive, the social character of the problem, and the direct, but variable importance of time lags should influence what is to be considered fair. Economists cannot answer such questions – they can neither disprove nor embrace certain conclusions – without taking ethical stands. Moreover, what may be believed to be ethically invariant, given the abstractions of the economic model, may not be so across real cases or settings. As economists, we should thus be much more careful when claiming a certain solution to be an "efficient resource allocation." Too many normatively contested issues are concealed within that phrase.

Regulation, Transaction Costs, and Fairness

The final step in formulating an environmental policy will be to choose policy instruments. We also face important problems of fairness here. To start with, we should observe that so-called efficient instruments such as effluent taxes do not necessarily create a strong link between the level of environmental damage and the costs imposed upon those designated as being responsible. The Pigovian solution is to set the tax equal to the marginal cost of damage in the optimum, and to require the emitter to pay the tax. Since the marginal environmental costs tell little about total costs, situations may emerge in which the tax burden is highest for the polluters that cause the lowest total environmental costs. While still technically efficient, this outcome may conflict with widely shared interpretations of what is fair. The effect is further accentuated if people interpret environmental taxes as punishments rather than as incentives.

More important ethical paradoxes surface if we take into consideration that what constitutes a Pareto-relevant externality depends on the level of transaction costs (Dahlman, 1979). When we include transaction costs, a move from effluent taxes to input taxes may become attractive, because doing so could reduce total costs. However, the distributional effects of this move may be significant, as will be demonstrated below.

Analysis of the physical characteristics of environmental problems helps to reason out the significance of distributive consequences. Environmental problems are foremost a result of the size and form of humanly influenced cycles of matter. Therefore, it appears to be highly relevant to focus on the input side of human activities. All matter that is taken into the economy will sooner or later be returned to the environment – in the form of wastes or pollutants. The number of input points – extraction activities – is much lower than the number of emission points in extraction, processing, and final consumption. Processing and consumption involve a large number of agents. Moreover, measuring inputs is usually much easier than measuring emissions. Thus transaction costs of input taxes will be lower. However, the reduction of costs has to be contrasted with the loss of precision: input taxes secure equality between gains and losses at the margin only in some cases. Still, there is probably a wide range of cases in which input regulations would be favorable in terms of total costs (Vatn, 1998). Moving the focus from emissions to inputs may thus change many environmental problems from being Pareto-irrelevant to Pareto-relevant.

The extraction of coal and oil and the subsequent release of CO_2 provides a good example. Since almost all emissions end up as CO_2 and are perfectly mixed, input regulations are likely to be as precise as emission regulations. However, transaction costs will be much lower in the first case. Just think about the number of cars and other emission points in the world. Many sources of carbon dioxide emissions would be deemed Pareto-irrelevant under an emission regulation regime, but found to be Pareto-relevant if regulation were to focus on inputs. This indicates that what appears to be Pareto-relevant depends on the chosen policy regime.

Input taxes usually have distributional effects that differ from effluent or emission taxes, even in the special case in which the instruments are equally precise. Stevens (1988) shows that if emissions are convex in inputs, more fees will have to be collected in the case of an input tax as compared with a charge on emissions. When emissions are concave in inputs, the conclusion is the opposite. In the case in which the relationships are linear, the two have equal distributional effects. To reach this conclusion, Stevens assumes that environmental damages are proportional to emissions. The dominant view is that the marginal damage increases together with emissions. The environmental cost of an extra unit of emissions is lower at low emission levels than at high emission levels. Thus, taking nature's cleaning capacity into consideration, most relationships between inputs and environmental damages will be convex.

Furthermore, not all uses of a certain compound will result in negative environmental effects. Taxing inputs will thus also hit environmentally "innocent" uses. Still, this may be the least-cost solution when transaction costs are taken into consideration. Vatn et al. (1997) illustrate this in a study of agricultural nitrate emissions. They show how variations in nature's cleaning capacity and the effects of different uses of inputs make the link between inputs and environmental damages weak. Despite this lack of precision, input taxes seem to be the low-cost solution from a societal point of view. The problem is that the input tax has significant distributive effects. To attain the same environmental effect, the input tax creates a 2–3 times higher tax burden than would a tax determined on the basis of damages (see Vatn et al., 1997).

This illustrates the problems related to the "tradeoff" between efficiency and fairness. One may ask whether it is justifiable to take the characteristics of material cycles and transaction costs into consideration when deciding the size of the "incentives" or burdens to be imposed upon the regulated agents. According to the logic of criminal law, if – in a case in which one knows that some are innocent – the only option is to punish either all or none, then none will be punished. The logic of criminal law may not be perfectly applicable here, but it still raises an important issue for consideration.

For economists, the lesson is not to stop searching for less costly options. Yet it must be acknowledged that both the level and the *distribution* of costs are influenced by institutional solutions. Therefore, there is a need to systematically analyze the distributive consequences of institutional (policy) alternatives. There is also a need to search for policy alternatives that can strike a balance between cost minimization and fair distributive consequences. For example, the use of input taxes could be accompanied by different kinds of compensation.

Conclusions

This chapter has examined the effect of interdependencies on the relationships between efficiency and fairness. The conventional neoclassical economic model establishes a demarcation line between the two, by assuming the independence of individual agents' goods, costs, and preferences. This simplification is obtained by describing the physical world from the commodity perspective and by restricting the parts of the world that are focused on by, for example, exempting important institutional issues.

This analytic convention produces problems at many levels of analysis. The problems are especially evident in environmental economics. There are two reasons for this. First, environmental problems are characterized by physical interdependencies and their direct effects via time lags, irreversibilities, and transaction costs. Second, these physical interdependencies contribute to the interdependence of agents' values and preferences.

The attributes of the ecological system complicate the search for an efficiency rule. The character of environmental problems means that environmental economics is constantly confronted with situations in which cost minimization and fairness considerations conflict with each other. For example, while transaction cost arguments often favor input regulations, they may burden "innocent" agents and have problematic distributive effects.

Even more importantly, due to the complex dynamics of economic and environmental systems, rights usually have to be defined *ex post*. This leaves neoclassical economists with the potential Pareto improvement rule as the only tool with which to deduce what it is efficient to do. Therefore, we face the problem of whether evaluations should be made on the basis of the old or the new rights structure. Economics cannot answer this question. Yet the use of the existing right structure has become a convention – a convention that has problematic allocative and distributive consequences.

The asymmetry of incentives for investments in markets and in the environment accentuates the foregoing problems. The "value of nature" reflects to a large degree the higher relative value of investments in markets for the individual when nature is presumed to be a free good. To break this logic, society has to make explicit decisions about which *development path* it wants to follow concerning the balance between market and environmental goods.

We have also demonstrated that the preferences of one individual influence the opportunities that others will have. This raises profound normative issues concerning the direction of economic development. Not only do we have to agree on what nature is and what should be conserved or developed, but one person's willingness or unwillingness to support environmental conservation – his or her choice of environmental preferences – will directly influence what others can consume.

An ethical standpoint that is characterized by individualism, welfarism, and universalism seems to face difficulties in observing and handling the structures and interconnections that are so crucial in choices concerning the environment. Simply understanding these shortcomings better would be a step toward improved environmental economics. We should abstain from advocating procedures that are built on a distinction between what is efficient and what is fair. Instead, we should favor a systematic treatment of interdependencies between rights, costs, and distributive effects. They simply cannot be isolated from each other.

References

Barry, B. 1989: *Theories of Justice. A Treatise on Social Justice*, Vol. I. Berkeley: University of California Press.

Baumol, W. J. and Oates, W. E. 1988: *The Theory of Environmental Policy*. Cambridge: Cambridge University Press.

Beck, U. 1992: *Risk Society. Towards a New Modernity*. London: Sage.

Becker, G. S. 1976: *The Economic Approach to Human Behavior*. Chicago: University of Chicago Press.

Bromley, D. W. 1989: *Economic Interests and Institutions. The Conceptual Foundations of Public Policy*. Oxford: Blackwell.

Bromley, D. W. 1990: The ideology of efficiency: searching for a theory of policy analysis. *Journal of Environmental Economics and Management*, 19, 86–107.

Calabresi, G. 1991: The pointlessness of Pareto: carrying Coase further. *Yale Law Journal*, 100, 1211–37.

Ciriacy-Wantrup, S. V. 1968: *Resource Conservation: Economics and Policies*. Berkeley: University of California Press.

Coase, R. H. 1960: The problem of social cost. *Journal of Law and Economics*, 3, 1–44.

Dahlman, C. J. 1979: The problem of externality. *Journal of Law and Economics*, 22, 141–62.

Elster, J. 1983: *Sour Grapes: Studies in the Subversion of Rationality*. Cambridge: Cambridge University Press.

Elster, J. 1992: *Local Justice. How Institutions Allocate Scarce Goods and Necessary Burdens*. New York: Russell Sage Foundation.

Graves, J. and Reavy, D. 1996: *Global Environmental Change*. Harlow, Essex: Longman.

Griffin, R. C. 1995: On the meaning of economic efficiency in policy analysis. *Land Economics*, 71, 1–15.

Mishan, E. J. 1971: The postwar literature on externalities: an interpretative article. *Journal of Economic Literature*, 9, 1–28.

Mishan, E. J. 1975: *Cost–Benefit Analysis: An Informal Introduction*. London: George Allen & Unwin.

Mishan, E. J. 1980: How valid are economic evaluations of allocative changes? *Journal of Economic Issues*, 14, 143–61.

Nordhaus, W. D. 1993: Rolling the "DICE": an optimal transition path for controlling greenhouse gases. *Resources and Energy Economics*, 15, 27–50.

Norton, B. 1992: Sustainability, human welfare, and ecosystem health. *Environmental Values*, 1, 97–111.

Norton, B., Costanza, R., and Bishop, R. C. 1998: The evolution of preferences. Why "sovereign" preferences may not lead to sustainable policies and what to do about it. *Ecological Economics*, 24, 193–211.

Nöel, J. F. and O'Connor, M. 1998: Strong sustainability and critical natural capital. In S. Faucheux and M. O'Connor (eds.), *Valuation for Sustainable Development*. Cheltenham, UK: Edward Elgar, 75–98.

Ostrom, E. 1990: *Governing the Commons. The Evolution of Institutions for Collective Action*. Cambridge: Cambridge University Press.

Parfit, D. 1983: Energy policy and the further future: the identity problem. In D. MacLean and P. F. Brown (eds.), *Energy and the Future*. Totowa, NJ: Rowman & Littlefield, 179–97.

Perrings, C. 1987: *Economy and the Environment*. Cambridge: Cambridge University Press.

Perrings, C. 1997: Ecological resilience in the sustainability of economic development. In *Economics of Ecological Resources. Selected Essays*. Cheltenham, UK: Edward Elgar, 45–63.

Pezzey, J. 1988: Market mechanism of pollution control: "polluter pays," economic and practical aspects. In R. K. Turner (ed.), *Sustainable Environmental Management: Principles and Practice*. London: London Belhaven Press, 190–241.

Randall, A. 1988: What mainstream economists have to say about the value of biodiversity. In O. E. Wilson (ed.), *Biodiversity*. Washington, DC: National Academy Press.

Rawls, J. A. 1971: *A Theory of Justice*. Cambridge, MA: The Belknap Press of Harvard University Press.

Schmid, A. A. 1987: *Property, Power, and Public Choice. An Inquiry into Law and Economics*. New York: Praeger.

Sen, A. 1979: Utilitarianism and welfarism. *Journal of Philosophy*, 76, 463–86.

Sen, A. 1987: *On Ethics and Economics*. Oxford: Clarendon Press.

Spash, C. L. 1993: Economics, ethics, and long-term environmental damages. *Environmental Ethics*, 15, 117–32.

Stevens, B. K. 1988: Fiscal implications of effluent charges and input taxes. *Journal of Environmental Economics and Management*, 15, 285–96.

Toman, M. A. 1994: Economics and "sustainability": balancing trade-offs and imperatives. *Land Economics*, 70, 399–413.

Vatn, A. 1998: Input vs. emission taxes. Environmental taxes in a mass balance and transactions cost perspective. *Land Economics*, 74(4), 514–25.

Vatn, A. 2000: The environment as a commodity. *Environmental Values*, 9, 493–509.

Vatn, A. and Bromley, D. 1997: Externalities – a market model failure. *Journal of Environmental and Resource Economics*, 9, 135–51.

Vatn, A., Bakken, L. R., Botterweg, P., Lundeby, H., Romstad, E., Rørstad, P. K., and Vold, A. 1997. Regulating nonpoint-source pollution from agriculture – an integrated modelling analysis. *European Review of Agricultural Economics*, 26, 207–29.

Walzer, M., 1983. *Spheres of Justice. A Defence of Pluralism and Equality*. New York: Basic Books.

11

Trading with the Enemy? Examining North–South Perspectives in the Climate Change Debate

Bhaskar Vira[1]

Textbooks on environmental economics extol the cost savings that can be made by using "incentive-based instruments" in environmental policy. For example, models demonstrate that the costs of attaining any reduction in pollution can be minimized by using instruments such as marketable permits. Indeed, trading systems have increasingly been used in domestic environmental policy since the 1970s, most extensively in the United States in the Acid Rain Program under the Clean Air Act amendments of 1990. The extension of these principles to international environmental policy has been a controversial aspect of the ongoing negotiations around the United Nations Framework Convention on Climate Change. In particular, proposals for emissions trading have been received with varying degrees of enthusiasm, despite their seemingly uncontroversial economic logic.

One obvious difference between domestic and international policy is the need for consensus among transacting parties (usually sovereign nations) in the latter context. The debate about trading on emissions globally is one that exemplifies the difficulties of achieving such a consensus, especially with regard to the initial allocation of emission rights to different countries. There is greater agreement about the economic rationale for emissions trading: given that marginal abatement costs vary between polluters, a system of marketable permits shifts the burden of actual abatement to low-cost locations. This can be shown to minimize the costs of reducing the global emissions of greenhouse gases (GHGs). Furthermore, such a mechanism is in the interests of both the buyers and sellers of permits. Buyers are able to meet their emissions reduction commitments at

[1] I am grateful to Dan Bromley and Jouni Paavola for their insightful comments, and to Harriet Bulkeley, who read an earlier version. The usual disclaimers apply.

costs that are lower than the domestic costs of compliance. Sellers anticipate receiving financial and technological transfers in return for the sale of permits.

The literature[2] suggests that the cost-effectiveness and efficiency of emissions trading is unrelated to issues of equity and distribution. It is also believed that the difficulty in implementing emissions trading for GHGs lies in the lack of global agreement on equity, particularly due to differences between the perceptions of the developed world and developing countries. These differences arise, in part, because the initial allocation of permits has distributional consequences. Sovereign countries clearly prefer schemes that maximize their own allocations, and the rents that derive therefrom, and seek to promote general rules that do not excessively damage their own interests. Negotiators are concerned about the burden of emissions reductions, and that this is shared equitably between countries. The transparency and fairness of the process by which allocation criteria are chosen in the international arena, and the manner in which divergent views are represented in the negotiation process, are also the subject of scrutiny. Governments, as well as nongovernmental organizations, are increasingly sensitive to the need to ensure that the process of consultation and decision-making is seen to be procedurally equitable (Banuri et al., 1996). This has an impact on discussions around the United Nations Framework Convention on Climate Change (see UNFCCC, 1997). These equity concerns are clearly central to achieving global agreement on responses to climate change, but it is widely believed that such issues are unrelated to the efficiency of emissions trading systems.

This chapter scrutinizes the economic logic behind emissions trading, and argues that the separation of equity and efficiency issues may be misguided. The first section outlines the usual economic argument for emissions trading. Conventional analyses that assume that the initial distribution of permits has no impact on abatement cost functions are then examined. In the standard economic model, the costs of abatement are known to firms, who choose their preferred level of pollution reduction on the basis of these (internal) costs as well as the price of permits. The marginal abatement cost function is derived purely from the technical options that are available to firms, about which the firm is assumed to have perfect knowledge. This suggests that the menu of technological alternatives is also independent of the initial distribution of permits. Such assumptions are problematic when analyzing options for reducing human impacts on climate change. The supply of technical solutions may be strongly influenced by the incentives to search for these alternatives. The section argues that the availability of opportunities for trade and the extent of the market for emissions permits both influence incentives to search for low-cost abatement options. Thus, the evolution and diffusion of technical options are not independent of distributional issues: initial allocation of permits may influence abatement costs. There is a link between equity and efficiency, and they cannot be treated separately in international negotiations.

---

[2]   There is a growing literature in this field. For a flavor of some of the debates, the interested reader could turn to Cline (1992), OECD (1992), Schelling (1992), Nordhaus (1993), Rose and Stevens (1993), Solomon (1995), Bush and Harvey (1997), Rose et al. (1998), Tietenberg (1998), or Toth (1999).

If abatement costs are influenced by the initial allocation of permits, their allocation has an impact on the *efficiency* of emissions trading, as well as on the distribution of burdens between different countries. The specific configuration of initial entitlements influences the size of efficiency gains from emission trading. While the textbook models predict significant efficiency gains, they usually fail to discuss the institutional realities of implementing such schemes. The second section examines four issues that can reduce the efficiency gains from global emission trading. These include: (1) additional costs for uncompensated third parties, whose interests are ignored in bilateral negotiations; (2) underestimation of the costs of adopting new technology, and the opportunity costs of mitigating or reducing emissions; (3) the possibility of strategic behavior in bilateral transactions and the monitoring costs associated with such transactions; and (4) the extent of transaction costs inherent in a global trading system. These additional costs provide reasons to be less sanguine about the efficiency gains from global trading on greenhouse gas emissions.

The final section concludes with some thoughts about the implications of the arguments presented in this chapter for international negotiations over responses to climate change. The analysis suggests that negotiators cannot treat the efficiency and distributive effects of global emissions trading separately. To suggest that we should implement the most efficient solution, and worry about equity as a secondary (and inevitably more complex) problem, is to deny the established theoretical link between these issues. Furthermore, if the efficiency gains from emissions trading are exaggerated, there may be reason to question the conventional wisdom about the allocative superiority of such schemes.

The Economic Case for International Emissions Trading

The economic argument for emission trading rests on the observation that firms (and, by extension, countries) face varying costs of adopting emission reduction schemes. Any emission reduction target can be achieved at least cost if the marginal cost of abatement is equalized across all sources. A perfectly functioning trading system secures such a result because each source chooses a level of control that equates its marginal abatement costs with the price of a permit. Firms are assumed to be price-takers, and cost minimization by all firms ensures that the marginal costs of abatement are equalized across all sources. Furthermore, emissions do not exceed the predetermined global limit, since the number of issued permits corresponds to the desired emissions reductions.

The variation in marginal abatement costs between sources is one of the factors that drives this result. If all sources were identical, it would be easy to design a control strategy that could be applied universally to achieve a desired level of pollution reduction. The problem is that sources usually *do* vary in terms of abatement cost functions, *and* the regulator generally does not have this information (which is private to firms). With emissions trading, firms choose between the purchase (or sale) of permits and the extent to which they reduce their own emissions. The regulator's role is reduced to deciding on an initial allocation of permits, and allowing the market to function smoothly. The system works in a cost-effective way despite the regulator's lack of complete information.

In the case of reducing GHGs, it is generally believed that abatement costs are lower in parts of the developing world. The introduction of global emissions trading ensures that these low-cost options are exploited, since they have a market value even if they are not necessary to meet a country's own commitments for GHG reduction. As long as the costs of reductions differ in the North and the South, emissions trading offers mutually beneficial transactions that reduce the costs of achieving global targets. The buyers of permits would economize on the costs of meeting their own reduction targets: the market price of emission credits would be lower than abatement cost in the absence of trading opportunities. The sellers would receive payments that fully finance abatement activities and may also leave a surplus. This suggests that an equilibrium with emissions trading is Pareto-better than one with no trade.

A global trading system for emission permits implies substantial financial flows from the North to the South. This is considered to be in line with global equity. However, the debate gets complicated here. The initial allocation of permits has an impact on the volume of trade, and hence on the extent of global transfers.[3] Developing countries often make a case for receiving a higher quota of pollution permits. They argue that developed countries are responsible for much of the climate change problem, because of the cumulative impact of their historical and current emissions on radiative forcing. It is sometimes suggested that the allocation of permits should punish those nations that are most responsible for the human impact on global climate change. Furthermore, if all humans are seen to have an equal right to environmental space, a population-based allocation of emission permits would seem to be equitable. This would bias allocation of permits in favour of the South, since countries in the South have large populations. On the other hand, developed countries sometimes argue that developing countries will increasingly contribute to global warming in the future, and that their unconstrained growth could have catastrophic implications for emission levels. This reasoning assigns a greater responsibility to the developing world because of their *potential* contribution to climate change. In general, the North favors allocations that recognize current emission levels as a baseline against which future permit allocations are determined (or "grandfathering"). Each of these allocations has some merit; the difficulty arises because they are mutually incompatible as rules for permit allocation.

While the initial allocation of emission permits is seen to influence the distribution of costs and benefits between different countries, it is not recognized as affecting the efficiency or cost-effectiveness of emissions trading. It is suggested that we can separate considerations of efficiency and equity. The argument is made explicitly in a series of influential papers by Rose and his collaborators. For

---

[3] The North (the developed world) and the South (the developing world) are treated below as two homogenous groups. There *are* important differences between the countries in the North and the South, but this crude distinction is adequate for the purposes of the present analysis.

instance, Rose and Stevens write that "... the efficiency and equity aspects of tradable entitlements are generally separable ..." (1996: 55). They claim that "As predicted by the Coase theorem, no matter how entitlements are assigned, there is a single, least-cost mix of abatement levels between countries. Put another way, this means that post-trading abatement costs will always be the same for any given cost-effectiveness target no matter what the initial allocation of entitlements" (Rose and Stevens, 1996: 64–5). Rose et al. (1998: 26) suggest that "... a marketable permits scheme will be cost effective irrespective of how the permits are distributed." A footnote seeks to demonstrate why this is a reasonable claim "... the Coase Theorem means that efficient abatement of CO_2 is independent of the distribution of a fixed global supply of emission permits among countries, assuming the absence of significant transaction costs and income effects. We believe these assumptions are reasonable" (Rose et al., 1998: 47).

One implication of such reasoning is that the normative discussion of what constitutes a just initial allocation of entitlements in the context of GHG emissions can be kept distinct from analyses that demonstrate the global efficiency gains from emissions trading. The next section suggests that such a separation may not be theoretically valid. The distribution of permits may influence technological change and thus have a dynamic effect on efficiency. This could have significant implications for international negotiations. Disagreements over permit allocations would need to be resolved *before* evaluating the relative efficiency of such a response strategy.

Induced Technical Change and the Initial Allocation of Emissions Permits

The analytic contributions made by Rose and his colleagues provide important insights into the implications of alternative equity criteria for the allocation of initial permits to GHG emissions. Central to their arguments is the belief that there is a *unique* post-trade equilibrium in which GHG reductions are undertaken in the lowest-cost locations, and that this outcome is achieved regardless of the initial allocation of permits. The model specified in Rose et al. (1998: 33) makes the key assumption that drives this claim explicit: that the slope parameter of the abatement cost function for any country is given *exogenously*. Thus, the marginal abatement cost for each location is a technical, externally given parameter that is unaffected by the initial allocation of permits. Since the range of abatement solutions does not vary with respect to the original allocation, it follows that an efficient market will ensure that the least-cost options are always chosen, regardless of the distribution of initial entitlements.

The uniqueness of a post-trade equilibrium may disappear if the initial allocation influences mitigation costs. This link can be established by examining the dynamics of technological change. It has been demonstrated that technological development responds to market conditions (Arrow, 1962; Arthur, 1994). Costs are influenced by learning-by-doing and research and development is affected by expectations about future market opportunities. The size of the market is one significant factor, since it "induces investment in fixed cost (or other increasing returns) technologies" (Ades and Glaeser, 1999: 1025). In the context of climate change mitigation, an increasing number of studies suggest that technological mitigation solutions are not exogenous (see Grubb et al., 1995; Dowlatabadi, 1998; Goulder and Schneider, 1999; Newell et al., 1999; Grubb et al., 2000). Technological innovation and change are influenced by external pressures, which may arise because of policy changes or market opportunities, for example. The literature on induced technological change suggests that the abatement cost function is *not* invariant to the initial allocation of emission permits if the initial allocation influences the size of the market.

Consider a situation in which trading is not permitted, and some countries ("the South") have no binding commitments to reduce their emissions of GHGs. In such countries, opportunities to reduce emissions may exist that are considerably cheaper than those available in countries that do have binding commitments ("the North"). Opportunities in the South would remain unexploited if trade were not permitted. It is also possible that some of these opportunities would not be discovered at all, because governments – as well as the private sector in the South – would have no incentive to examine the nature of their operations or to invest in research and development in a no-trade situation.

Even if trade were to be permitted, the market value of tradable emissions reductions would have an impact on incentives to discover these options. The strength of such incentives may depend on the estimated size of the market for emissions permits. If the volume of trade were expected to be high, there would be a greater incentive to look at production to identify potential sources of emission credits. If the market is expected to be thin, perceived returns may not justify the search and research and development costs. This may be particularly true for locations that do not expect to be major players in the market, either because of their relative cost structures, or because of the expected volumes of tradable surpluses.

The size of the market, and the volume of trade, are affected by the initial allocation of permits. Rose et al. (1998) demonstrate this by simulating outcomes that may result from a number of plausible alternative criteria for the initial allocation of emission rights. The criteria used are "sovereignty" (a version of grandfathering), "egalitarian" (permits are allocated according to projected population levels), "horizontal" (net costs are assumed to be 1 percent of GDP for all countries), "vertical" (permits are allocated progressively according to per capita GDP), and "consensus" (equal weight is given to population and GDP). Table 11.1 is derived from the results reported in Rose et al. (1998), and illustrates the value of trade that would emerge under these alternatives.[4]

_____

[4]   As these are value figures, they represent an estimation of the physical volume of ade multiplied by the price of permits. Given that the global price of permits emerges as invariant to the initial distribution (Rose et al., 1998: 34), differences reported in table 11.1 can be seen to reflect differences in the volume of trade that would occur under alternative initial allocations of emissions permits.

Table 11.1 The value of trade under different rules of initial allocation

Equity criterion	Value of trade (in billions of 1990 US dollars)		
	2005	2020	2035
Sovereignty	5.1	9.4	9.0
Egalitarian	146.9	176.8	141.5
Horizontal	9.6	16.2	13.6
Vertical	17.6	28.0	15.4
Consensus	59.4	77.3	69.8

Source: Rose et al. (1998)

The simulations suggest that the extent of the market does vary considerably with the choice of allocative criterion, with expectedly wide divergence between the sovereignty rule and the egalitarian rule. The differences between the other alternatives are also significant, widening over the middle-range forecast and then closing in the long term as technologies become more convergent. These results confirm the intuition that the volume of trade will be larger if greater allocations of permits are made to developing countries. This would be the case under the population-based egalitarian criterion.

The size of the market has different impacts on the pace of technological change in countries that buy and sell emission permits. Countries with high abatement costs (buyers) would be expected to accomplish their GHG reductions by domestic means if trade is not permitted. In this limiting case, firms would have incentives to search for new technologies in order to lower the domestic costs of compliance. Once trade is permitted, there are fewer incentives for domestic action and research on them, since less expensive options would be available internationally. Grubb recognizes such disincentives, and warns that the lack of substantive domestic action by buyers "will be seen by developing countries as violating the spirit of the Kyoto agreement and the principle of the Convention itself" (Grubb, 1998: 142).

On the other hand, sellers are likely to be countries with low mitigation costs. For them, the prospects of trade open up new market opportunities. The potential for emissions trading would provide positive incentives for the identification of new emission reduction opportunities and technological change in these countries. The impact of trading on technological change at a global level will depend on the extent to which reduced innovation among buyers is balanced by the faster pace of technological progress among sellers. The overall pace of technological change will depend on whether the sources of such change are concentrated predominantly in buyer- or seller-countries, and the capacity for technological change in each location. If buyers are mainly in the North and sellers in the South, the result may be slower overall technological change, because of the greater short-term capacity for technological change in the advanced industrial nations. The associated higher costs of global adaptation and mitigation would then need to be offset by the "efficiency" gains that are presumed to follow from allowing trade in emission credits.

This analysis suggests that the initial allocation of emissions permits has an impact on the size of the market and the menu of mitigation options. The final outcome of trading is thus not independent of the initial allocation of permits. Indeed, for any given allocation, there is a set of mitigation possibilities (with an associated set of technological options), each of which could be seen as "efficient" with respect to the particular initial distribution of entitlements. If the initial allocation changes, an alternative technological trajectory would emerge, with associated impacts on the costs of emissions reduction. The implications of such reasoning are straightforward – the choice over initial allocation of permits affects not only the extent of financial transfers between North and South, but also the global costs of responding to climate change. To restate, the debate over initial entitlements is not simply one in which sovereign states are pursuing their self-interest – it also has significant implications for the global costs of reducing greenhouse gas emissions.

The next section introduces further complications to the textbook treatment of these issues. The received wisdom is that trade is always better than no trade, since any emissions reduction target can be achieved at the lowest possible cost under a trading system. The next section suggests that allowing trade may compromise overall emission reductions or not be the most cost-effective policy response. If these are valid qualifications, there may be reason to rethink current proposals for introducing "flexible" trading mechanisms to reduce global greenhouse gases, as some (particularly Southern) commentators have been arguing.

Is Some Trade Always Better Than No Trade?

The cost-effectiveness of emissions trading relies on the existence of a difference in marginal abatement costs between different locations. Mutually beneficial trading opportunities exist in this situation: high-cost locations can purchase emission credits from low-cost locations at a price that at least covers the cost of mitigation. Global cost-effectiveness is achieved if trading reduces global emission control costs, net of all transaction and opportunity costs. As long as there are net benefits from trade, a policy that permits emission trading is preferable to one that precludes it. However, these net benefits may be exaggerated, and allowing trade may actually reduce global mitigation efforts. If this is true, a regime that does not allow trade may be preferable to one that does.

A factor that is sometimes neglected in calculations is the impact that trade has on parties that are not represented in the transaction. While trade may be beneficial to the transacting parties, other parties may suffer from unaccounted for and noncompensated effects or externalities. For instance, Jepman and Munasinghe suggest that "a forestry project may be effective as a carbon store, but may force poor landless peasants to migrate to less productive areas" (1998: 296). Forestry projects have been seen as important sinks for GHGs, and potential sources of emission credits. To continue with the example of Jepman and Munasinghe, it is unlikely that the poor landless peasants are represented in international negotiations over emissions credits and the price at which the credits are sold. If such groups are also voiceless in domestic politics in the host country, negative impacts of mitigation projects on their well-being may simply be ignored. The existence of these third-party effects suggest that the opportunity costs of mitigation projects should be considered in the widest possible way. This may mean that the benefits from trade are lower than is usually thought.

Another type of externality may arise because of the impact of mitigation projects upon developmental alternatives in host countries. Forestry projects may be problematic, because they may exclude other land uses and compromise existing livelihood strategies (Cullet and Kameri-Mbote, 1998). Other projects may involve the transfer of leading edge technology. In this case, host countries need to expend resources to develop the capacity to work with such technologies. The expenditures may crowd out domestic spending in other sectors and have dynamic effects on developmental trajectories in these areas.

The extent to which trading can reduce global compliance costs will also be influenced by transaction costs. Rose et al. argue that transaction costs are likely to be "an infinitesimal fraction of the tens of billions of dollars of projected permit transactions" (1998: 47). However, Jepman and Munasinghe (1998: 305) report that there is "anecdotal evidence with respect to some US tradable permit schemes (mainly dealing with local pollution) that transaction costs can be considerable." This is an unresolved issue that is of considerable importance for the potential gains of emissions trading. If the identification of trading options, establishment of monitoring mechanisms, and enforcement of any penalties have setup or fixed costs, then the relative transaction costs will decrease as the volume of trade increases. This suggests that the level of transaction costs may be related to the initial allocation of permits, which determines the size of the market.

It is also difficult to monitor behavior after the introduction of a trading regime. For instance, Wirl, Huber, and Walker (1998: 205) argue that "the participants (the developing and the industrialised country) have an incentive to misreport, more precisely, to inflate the reported reduction of CO_2 emissions. As a consequence, hypothetical reductions of CO_2 emissions will be large while actual reductions are most likely to be small." This kind of strategic behavior arises because costs and current levels of performance are difficult to verify. Wirl et al. examine improving energy efficiency in the power sector. Substantial savings are expected in this area because of current low performance of coal-powered plants in Third World countries. Emission credits are generated when the investing country pays the costs of improving the plant's performance, financing a project that would not otherwise have been undertaken by the host country. Wirl et al. show how this may lead to an under-reporting of current performance by the host country. This is also compatible with the interests of the investing country, since they can claim larger emissions reduction credits. Actual emission reductions may thus be less than claimed. The possibility of strategic behavior under a bilateral trading system would thus compromise the global level of emissions reductions.[5]

This discussion suggests that the efficiency gains that would result from global emissions trading may be lower than expected in theoretical models. The desirability of trading is not altered if gains from trade are high and the efficiency losses are modest. However, if gains from trade are modest and efficiency losses significant, there may be reason to question that some trade is better than no trade. Trade should be restricted purely on efficiency grounds, if the net benefits of trade are insufficient to offset the costs of third-party effects, opportunity costs, costs of absorbing new technologies, and transaction costs. Furthermore, the possibility of strategic behavior may compromise global emission reduction targets. It may be preferable (and cheaper) to prohibit trade altogether, rather than creating an elaborate infrastructure to monitor and sanction such behavior.

These theoretical reflections suggest that there are two separate, but mutually reinforcing reasons, to be cautious about proposals that advocate emissions trading as an efficient response to the climate change. The first set of arguments demonstrates that endogenizing the process of technological change in predictive models challenges the separation of distributive issues from cost-effectiveness. There are also reasons to question the efficiency gains of emissions trading and to believe that trade outcomes may compromise overall global emission reduction targets. This chapter has so far not made direct reference to actual international negotiations over climate change, in order to avoid detracting from the theoretical discussion that is at the heart of its argument. The next section will contextualize the arguments with reference to the negotiations on the Kyoto Protocol to the United Nations Framework Convention on Climate Change (UNFCCC). The focus is on the difficulties of attempting to separate distributional issues from questions of allocative efficiency in the negotiations.

The Kyoto Protocol and Proposals for Global Emissions Trading

The Kyoto Protocol was concluded in December 1997 at the Third Conference of the Parties (CoP3) to the UNFCCC. Developed countries committed themselves to lower their emissions to at least 5.2 percent below the 1990 levels during the period 2008–12. The Protocol also contained provisions for flexible instruments to promote cost-effective ways of meeting commitments. Article 17 allows countries to meet their commitments through emissions trading, subject to the

---

[5] In their paper, Wirl et al. (1998) propose an incentive compatible mechanism that would address the issue of strategic behavior by modeling a three-player interaction between an implementing body, an investing country, and a host country. The implementing body sets a baseline level of emissions for the host country, and allows the investing country to claim credit for emissions that are reduced below this baseline. The results suggest that the need to deter strategic behavior results in significant efficiency losses and that host countries would receive lower transfers, the more efficient host countries benefiting disproportionately.

condition that "any such trading shall be supplemental to domestic actions." Another flexible mechanism is Joint Implementation, which allows Annex 1 countries to claim credits for cross-border investments in other Annex 1 countries (Article 6 of the Protocol). The Protocol also established the Clean Development Mechanism (Article 12), which allows Annex 1 countries to gain credits from similar activities with non-Annex 1 countries. The Protocol indicates that the Clean Development Mechanism (CDM) is intended "to assist Parties not included in Annex 1 in achieving sustainable development and in contributing to the ultimate objective of the Convention, and to assist Parties included in Annex 1 in achieving compliance with their quantified emission limitation and reduction commitments." Project activities in non-Annex 1 countries generate Certified Emissions Reductions (CERs) that can be used by Annex 1 countries to meet their commitments. The Conference of the Parties has the responsibility of supervising the Mechanism. Emissions reductions may be certified subject to: (a) the voluntary participation of each involved Party; (b) real, measurable and long-term benefits related to the mitigation of climate change; and (c) reductions in emissions that are additional to any that would occur in the absence of the certified project activity. The Protocol also specifies that "a share of the proceeds from certified project activities [would be] used to cover administrative expenses as well as to assist developing country Parties that are particularly vulnerable to the adverse effects of climate change to meet the costs of adaptation."

For most OECD countries, the CDM is a way of minimizing the costs of meeting their commitments agreed at Kyoto (Grubb et al., 1999). CDM is also expected to assist in the transfer of technology to the developing countries. It is worth examining the initial allocations of emissions permits that are implicit in the trading mechanisms of the Kyoto Protocol. Annex 1 countries have made commitments to reduce emissions to a level that is related to their 1990 emissions. This is a version of "grandfathering." Other countries have no binding commitments, which is why the process of certifying emissions reductions becomes important. This would be done, presumably, on a case-by-case basis. The key requirement is that emission savings arising from a CDM project must be additional to what would have happened otherwise. However, since what would have happened otherwise is unobservable, certification is speculative and subject to error. Grubb et al. (1999: 227–8) suggest that it may be possible to quantify "additionality" for some projects, but that "these are the exceptions" because "the future is uncertain and decision-makers are human." Grubb et al. (1999: 230) conclude by arguing that "real policy should not be based on the chimera of accurate quantification." They propose to focus on the actual ability of projects to achieve real reductions in emissions, rather than on the paper transactions that the CDM allows through the creation of CERs. The issue of strategic behavior raised by Wirl et al. (1998) is also relevant here. An inflated baseline would generate higher CERs for host countries, without accompanying increases in actual emission reductions.

The adoption of binding commitments by Annex 1 countries has generated a demand for emission credits. However, since other countries have made no commitments, the supply of credits has to be created artificially. The principle of additionality may appear sensible as an accounting procedure. Yet if emissions reductions are genuinely higher than would have occurred in the absence of certified project activity, important *moral* questions emerge. Projects that generate additional emission reductions necessarily alter development priorities in host nations. In order to qualify for a CER, a project has to *demonstrate* that it would not have been otherwise undertaken by the host company or country in response to evolving market conditions or changes in domestic environmental policy. Sometimes truly additional projects would lead to a win–win outcome, facilitating the adoption of much-needed new technology in the host country as well as generating credits for the investor. This would be likely when the host country is already interested in the project but is constrained from implementing it because of inadequate financial or technological resources. However, other projects (such as the diversion of land from other purposes for forestry projects) may have significant opportunity costs. It is not clear that the standard financial and technical transfers would provide adequate compensation in such cases. Since projects have to *prove* that they are not results of ongoing policy changes in the host, this process does change the host's developmental priorities and trajectories. If significant external costs are imposed by projects on some groups, it is important to ensure that these costs are incorporated into the prices at which emission credits are exchanged. Without such safeguards, projects that can demonstrate additionality may compromise hosts' national developmental objectives.

Ensuring adequate compensation is likely to be particularly difficult if the projects disadvantage groups that have no influence on domestic policy-making in the host country. Those who negotiate projects in the international arena may not represent the interests of poor and politically weak groups that are most impacted by the projects. If projects are implemented primarily to create a supply of CERs to reduce the cost of meeting commitments by the investing countries, and they negatively affect the livelihood of poor and weak groups (or foreclose particular development trajectories in the host country), they raise profound questions about international equity. It is impossible to discuss the cost reductions from meeting commitments through the CDM without first addressing these equity issues. What is efficient will be based on the adoption of specific projects, with associated distributional impacts. Equity and efficiency are intrinsically linked in these circumstances, and it is untenable to separate them.

Developing countries have expressed concerns for the prospect of unequal exchange of CERs under the Kyoto Protocol (Yamin, 1999). Although participation in the CDM is voluntary, bilateral bargaining will determine the terms of trade. CSE (1998: 4) argues that unless "the South [is] paid a 'fair' price – which accounts for its present and future needs – for its emissions ... CDM amounts to a global carbon scam and makes the sale of Manhattan for a few beads pale in comparison." The alternative would be to replace bilateral trading with a portfolio or multilateral approach. In this solution, the CDM would act as an intermediary, aggregating projects from the host countries and selling them on to investors.[6] Although the choice of the trading scheme seems to be largely a distributional issue (because it influences the terms on which CERs are exchanged between buyers and sellers), transaction costs will vary significantly between bilateral and multilateral trading systems. Grubb et al. (1999: 234) argue that the biggest risk of the multilateral model is that "separating foreign investor from host – and including concerns about equity, sustainable development criteria, etc. – could make a hugely bureaucratic and inefficient structure which would hardly operate as an international market mechanism at all." On the other hand, the transaction costs of identifying parties and projects and monitoring compliance are likely to be significant under the bilateral model. Both approaches will probably be used. They will influence not only the amount of financial transfers between North and South, but also the level of transaction costs and hence the global efficiency gains associated with emissions trading.[7]

One objection to emissions trading and to the provisions of the CDM in particular is that they allow developed countries to meet their Kyoto obligations without domestic action. Since the CDM allows developed countries to transfer some burden of reductions away from their own economies, Southern commentators see it as having been "designed to help the rich and not the poor" (CSE, 1998: 2). There is an important equity issue about the allocation of responsibility for global warming, one that is still unresolved because of different interpretations of such responsibilities. Differences prevail because some analyses examine the past contribution of different countries to radiative forcing, while others use projections of future emission levels as a basis for allocating responsibilities.[8] My analysis suggests that it is inappropriate to treat this issue as one which has purely distributional implications. If the induced technical change argument is valid, the ability to trade may have a negative effect on incentives for technological change in the developed world, since cheaper options may be available internationally. An equilibrium with trade would be associated with a different technological trajectory compared with one where trade was precluded. Allowing trading not only permits the North to meet some of its global obligations through nondomestic action, but also reduces incentives to invest in the development of new technology in the North. If the North drives the development of new technology, a post-trade equilibrium would be one with less rapid technological change, with associated impacts on dynamic efficiency.

---

[6] The recently launched Prototype Carbon Fund, set up by the World Bank, is an example of a multilateral mechanism that has been created to stimulate the market in emissions reductions (World Bank, 2000).

[7] The credibility of the intermediary is critical to the success of such multilateral approaches. A prominent Southern group, the Centre for Science and Environment in India, sees such schemes as a means of "securing the lowest cost options for the buyers" (CSE, 1998: 5). Furthermore, they point out that at least four multilateral agencies (UNEP, UNDP, World Bank, and UNCTAD) are competing for a role as brokers of developing-country interests. They suggest that the presence of these agencies may increase pressures on developing countries to participate in the emissions trading mechanisms, despite the lack of agreement about the equity of initial entitlements, or the legitimacy of trading itself.

[8] For a discussion of some of these issues, see Yamin (1999).

Other studies have sought to link the volume of emissions trading with global costs of emissions reduction. For instance, Hamwey and Baranzini (1999) have tried to estimate the size of the GHG offset market that is likely to emerge from commitments made by Annex B countries under the Kyoto Protocol. They (1999: 125) recognize that "... abatement costs ... vary considerably depending on the size of the market." In support of the argument forwarded here, they point out that the full spectrum of supply-side options has not yet been identified. However, their analysis is based on the expectation that marginal abatement costs would rise as abatement efforts increased. As more international options are demanded, progressively more costly options would be brought to the market.[9] The argument based on induced technical change is different, since it suggests that the size of the market affects not only the *equilibrium point* on the global abatement cost function, but also the *slope* of the marginal abatement cost function itself. Hamwey and Baranzini work with the assumption of a single global marginal abatement cost function that is invariant to the initial distribution. This chapter suggests that the initial distribution would affect the evolution of technology, and hence the global costs of meeting agreed emissions reduction targets.

In the present context, efficiency should be taken to refer not just to the costs of achieving an agreed target for reducing emissions of GHGs, but also to ensuring that this target is actually achieved. In his discussion of implementation of emissions trading under the Kyoto Protocol, Grubb (1998: 141) proposes that "the trading system should be implemented in ways that ensure that no trade of 'assigned amounts' can lead to collective emissions being higher than in the absence of that trade." This, he says, ensures "emissions conservation: trading should not under any circumstances be a vehicle for weakening the overall degree of limitation as compared with the situation in the absence of those trades." One reason why this principle may be violated is the strategic behavior analyzed by Wirl et al. (1998): host countries have incentives to inflate the contribution made by CDM projects, or to fail to comply with agreed reductions. Another reason may be the allocation of "assigned amounts" that exceed the country's projected emissions under "business as usual," which are referred to as "hot air." Grubb (1998) argues that allocations for Russia and the Ukraine may make available such surpluses to other countries to weaken their own domestic commitments. In this case, the initial allocation of entitlements and the possibility of trade influence the actual global emissions reductions, and hence the effectiveness of emissions trading. Hot air would not be a problem for global GHG reductions in the absence of emissions trading. The hot air problem can become more acute if developing countries are forced to adopt baselines for their projected emissions. It would be in their interests to adopt inflated baselines, and negotiators may see these as an acceptable price to pay to induce the developing

---

[9]   According to Hamwey and Baranzini (1999), if 50 percent of agreed commitments are made through international flexible mechanisms, the equilibrium price of permits is around $15 per tonne of carbon. The price rises to $20 per tonne of carbon if 75 percent of commitments are met in this way. This reflects simply the increasing *scarcity* of cheap mitigation options.

countries to take on voluntary commitments (Grubb et al. 1999: 262–5). This additional hot air would further dilute the current targets of developed countries. Again, the initial allocation is not purely a distributional issue: it directly influences the environmental effectiveness of equilibrium outcome.

This discussion of negotiations around the UNFCCC demonstrates that questions of distributional equity and allocational efficiency are very closely linked. It is disingenuous to argue that concrete action based on the efficiency implications of emissions trading can proceed without directly confronting the *prior* question of initial entitlements. Therefore, the insistence (particularly of Southern groups) that trading mechanisms should not be implemented without achieving global agreement on entitlements has strong theoretical support. Furthermore, the introduction of trading mechanisms in the absence of such discussion presumes that the existing global distribution of entitlements is widely accepted as equitable. This is clearly not true.

Conclusion

Global negotiations around the UNFCCC have primarily focused on minimizing the costs of meeting GHG reduction targets, neglecting or avoiding equity concerns (Shukla, 1999). The justification offered for this approach has been the claim that it is theoretically possible to separate efficiency and equity aspects of policy problems. This chapter has argued that the theoretical separation of distributive and allocative concerns is not valid. Negotiators in the international arena must accept the challenge of dealing with the contentious political issue of allocating initial entitlements at the same time as they debate mechanisms to minimize the costs of complying with agreed GHG reduction targets.

The chapter has challenged the received wisdom by examining the implications of modeling technological change as endogenous. These models suggest that the initial allocation of entitlements influences efficiency through an effect on the size of the market for emission permits. Furthermore, other factors such as external effects on third parties and opportunity costs in host countries also influence efficiency as well as the effectiveness of GHG reduction strategies. The principal conclusion is that it is not possible to treat efficiency and equity as distinct issues in international negotiations. Moreover, it is questionable whether trading *can* always deliver the target level of global emission reductions, and at lower cost overall.

This reasoning has ramifications for strategies for negotiators in discussions about instruments to implement GHG reductions. Southern objections to the introduction of emissions trading are often dismissed as merely being attempts to secure distributionally advantageous initial allocations. Yet the introduction of emissions trading despite failure to agree on international equity issues implies recognizing existing emissions as a basis for initial allocations. Such allocations favor the North and are considered detrimental to Southern interests. This chapter indicates that such an initial allocation of entitlements would also affect the efficiency gains of emissions trading and may also compromise the achievement of global emission reduction targets. The international process must address the distributional issues as a first priority, and this may also help secure wider participation in activities associated with the UNFCCC. The design of institutional arrangements to maximize the efficiency gains of emissions trading is sensible only after this conflict over initial allocations is resolved.

References

Ades, A. F. and Glaeser, E. L. 1999: Evidence on growth, increasing returns and the extent of the market. *Quarterly Journal of Economics*, 114, 1025–45.

Arrow, K. J. 1962: The economic implications of learning-by-doing. *Review of Economic Studies*, 29, 155–73.

Arthur, W. B. 1994: *Increasing Returns and Path Dependence in the Economy*. Ann Arbor, MI: University of Michigan Press.

Banuri, T., Maler, K.-G., Grubb, M., Jacobson, H. K., and Yamin, F. 1996: Equity and social considerations. In J. P. Bruce, H. Lee, and E. F. Haites (eds.), *Climate Change 1995: Economic and Social Dimensions of Climate Change*. Cambridge and Melbourne: Cambridge University Press for the Intergovernmental Panel on Climate Change, 79–124.

Bush, E. J. and Harvey, L. D. D. 1997: Joint implementation and the ultimate objective of the United Nations Framework Convention on Climate Change. *Global Environmental Change: Human and Policy Dimensions*, 7, 265–85.

Cline, W. R. 1992: *The Economics of Global Warming*. Washington, DC: Institute for International Economics.

CSE 1998: *The (UN) Clean Development Mechanism*. CSE dossier Fact Sheet 3 for the Fourth Session of the Conference of the Parties to the United Nations Framework Convention on Climate Change, New Delhi: Centre for Science and Environment. See also http://www.oneworld.org/cse/html/cmp/pdf/fact3.pdf

Cullet, P. and Kameri-Mbote, A. P. 1998: Joint implementation and forestry projects: conceptual and operational fallacies. *International Affairs*, 2, 393–408.

Dowlatabadi, H. 1998: Sensitivity of climate change mitigation estimates to assumptions about technical change. *Energy Economics*, 20, 473–93.

Goulder, L. H. and Schneider, S. H. 1999: Induced technical change and the attractiveness of CO_2 abatement policies. *Resource and Energy Economics*, 21, 211–53.

Grubb, M. 1998: International emissions trading under the Kyoto Protocol: core issues in implementation. *Review of European Community and International Environmental Law*, 7, 140–6.

Grubb, M., Chapuis, T., and Minh, H. D. 1995: The economics of changing course: implications of adaptability and inertia for optimal climate policy. *Energy Policy*, 23, 417–31.

Grubb, M., Vrolijk, C., and Brack, D. 1999: *The Kyoto Protocol: A Guide and Assessment*. London: The Royal Institute of International Affairs.

Grubb, M. and Kohler, J. 2000: *Induced Technical Change: Evidence and Implications for Energy-Environmental Modelling and Policy*. Report to the Organisation for Economic Cooperation and Development (OECD), Paris.

Hamwey, R. and Baranzini A. 1999: Sizing the global GHG offset market. *Energy Policy*, 27, 123–7.

Jepman, C. J. and Munasinghe, M. 1998: *Climate Change Policy: Facts, Issues and Analyses*. Cambridge: Cambridge University Press.

Newell, R. G., Jaffe, A. B., and Stavins, R. N. 1999: The induced innovation hypothesis and energy-saving technological change. *Quarterly Journal of Economics*, 114, 941–75.

Nordhaus, W. 1993: Reflections on the economics of climate change. *Journal of Economic Perspectives*, 7, 11–25.

OECD 1992: *Climate Change: Designing a Tradable Permit System*. Paris: OECD.

Rose, A. and Stevens, B. 1993: The efficiency and equity of marketable permits for CO_2 emissions. *Resource and Energy Economics*, 15, 117–46.

Rose, A. and Stevens, B. 1996: A tradable carbon entitlements approach to global warming policy: sustainable allocations. In P. H. May and R. S. da Motta (eds.), *Pricing the Planet: Economic Analysis for Sustainable Development*. New York: Columbia University Press.

Rose, A., Stevens, B. Edmonds, J., and Wis, M. 1998: International equity and differentiation in global warming policy: an application to tradable emission permits. *Environmental and Resource Economics*, 12, 25–51.

Schelling, T. C. 1992: Some economics of global warming. *American Economic Review*, 82, 1–14.

Shukla, P. R. 1999: Justice, equity and efficiency in climate change: a developing country perspective. In F. L. Toth (ed.), *Fair Weather: Equity Concerns in Climate Change*. London: Earthscan, 145–59.

Solomon, B. D. 1995: Global CO_2 emissions trading: early lessons from the US Acid Rain Program. *Climatic Change*, 30, 75–96.

Tietenberg, T. 1998: Economic analysis and climate change. *Environment and Development Economics*, 3, 402–5.

Toth, F. L. (ed.) 1999: *Fair Weather: Equity Concerns in Climate Change*. London: Earthscan.

UNFCCC 1997: Kyoto Protocol to the United Nations Framework Convention on Climate Change. See http://www.unfccc.org/resource/docs/convkp/kpeng.html

Wirl, F., Huber, C., and Walker, I. O. 1998: Joint Implementation: strategic reactions and possible remedies. *Environmental and Resource Economics*, 12, 203–24.

World Bank 2000: World Bank launches first-of-its-kind market-based Carbon Fund. World Bank News Release No: 2000/176/S.

Yamin, F. 1999: Equity, entitlements and property rights under the Kyoto Protocol: the shape of "things" to come. *Review of European Community and International Environmental Law*, 8, 265–74.

12

Social Costs and Sustainability

Martin O'Connor

It may be imagined, perhaps, that the law has only to declare and protect the right of every one to what he has himself produced, or acquired by the voluntary consent, fairly obtained, of those who produced it. But is there nothing recognized as property except what has been produced? Is there not the earth itself, its forests and waters, and all other natural riches, above and below the surface? These are the inheritance of the human race, and there must be regulations for the common enjoyment of it. What rights, and under what conditions, a person shall be allowed to exercise over any portion of this common inheritance cannot be left undecided. No function of government is less optional than the regulation of these things, or more completely involved in the idea of civilized society. (Mill, 1909 [1871]: 797)

The Problem of Social Choice, Revisited

Introduction

Should migratory bird and bear habitats be maintained? If mineral mining is a "must" for modern economies, how much community disruption, how much soil and water pollution is an acceptable price of progress? What criteria should be applied?

Amongst economists, our habit is to frame resource management analyses in terms of "supply" and "demand." On the supply side, the problem is to define the *frontiers of what is feasible* for the economy and, more especially, the tradeoffs (opportunity costs) imposed by the limits to what is feasible. On the demand side, the problem is to assess *what will be judged desirable* by members of the society.

In order to answer the economist's favourite question, "What is the highest-valued resource use?", the aim is to make the marginal costs of supply match with the marginal benefit to society (the social demand). But we run straight into, first, the problem of distribution and, second, the uncertainties of future supply.

In the case of a dam, or a motorway, or a forest exploitation scheme, or a fisheries management regime, or the introduction of genetically modified organisms, there

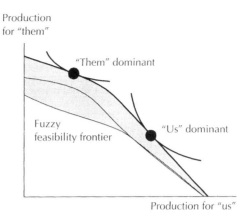

Figure 12.1 The fuzzy production possibilities frontier.

are a variety of uncertainties about outcomes, yet one thing is sure – there will be winners and losers and, moreover, the question of rights and duties is often in dispute.[1] The decisions and governance processes will in this respect involve what institutional economists Warren Samuels and Allan Schmid call sacrificial or moral choices (Samuels and Schmid, 1981).

Consider figure 12.1. Suppose that a simple economy can produce two types of goods, those favored by "us" (the horizontal axis) and those favored by "them" (the vertical axis). Limited resources mean a tradeoff between the two types of goods, represented by the heavy curve sloping from the upper left to the lower right. What is desirable depends on who you ask. If the economy is dominated by "us," the optimal mix of goods (suggested by the heavy spot toward the bottom right) is different from the optimal mix for an economy dominated by "them" (the heavy spot toward the upper left). The distribution of purchasing power is decisive for what will appear as the highest-value use of available resources, and this in turn depends on the prevailing conditions of access "rights" and constraining duties.

Traditional concerns with productive efficiency, resource discovery, and technological progress have put the emphasis on getting on to the frontier of feasibility and, going beyond, pushing out the frontier of possibilities. (In figure 12.1, this would be a movement "up and toward the right" of the whole frontier.) In the sustainability context, however, the core question is "What, and for whom?", and the focus must be on deciding *which feasible production* is to be distributed between "us" and "them."

For example, establishing a justification to build or not to build a dam depends on what forms of life and social relations shall be sustained, and what shall be foreclosed (cf., McCully, 1996). What is the basis for resolving the ownership,

---

[1] In the field of water resources and development, the recent work of the World Commission on Dams (2000) provides a rigorous and rich international documentation. See http://www.damsreport.org/

rights, duties, or wider distributional conflicts? How might considerations of desirability, which differ from "us" to "them" and from group to group within society, be expressed and reconciled? This is our first thematic starting-point (see the second subsection).

Our second starting-point relates to the uncertainties about outcomes and – what is worse – the certain (or uncertain) arrival of "bads" as a normal part of the possible outcomes. As suggested in figure 12.1, the future feasibility frontier fluctuates before our eyes. When we consider GMOs, nuclear fusion, cloning, portable telephone radiation, plastic bags, and other wonders of the modern world, then we see that – depending on the mind-set that you belong to – both Perpetual Purgatory and Paradise on Earth seem to be within the bounds of postulated feasibility. This is our second starting-point (third subsection), which leads in to the discussion of justifications for reflexive deliberation (fourth subsection).

Distribution: the impossible problem of social choice

In an attempt to get rid of vexatious "moral choice" issues associated with the distribution of wealth and the (re)distribution of sacrifice, economists have sometimes tried to separate out two levels: on top, like oil on troubled water, a political level that decides resource access rights and duties, and which is admitted to entail irreducible normative considerations; and underneath, a strictly economic level that is concerned with resource use efficiency and that is postulated to engage essentially "positive" (viz., objective, descriptive, value-neutral) analytic considerations.

However, it is well known that this attempted separation is really cheating, and its pursuit can lead on quite quickly to theoretical incoherence (not to say public disrepute). In the standard (neoclassical) economic theory, changing the rights structure may well change the "value maximizing" output mix, meaning that the answer to "What to do?" – What is the highest value resource use? – is determined essentially by distributional rather than efficiency considerations.[2]

Therefore, economists cannot avoid examining the basis for resolving a "just" or "best" or "socially optimal" distribution. The attempted axiomatization of this abstract social choice problem, as formulated by Kenneth Arrow, has led to an apparent impasse, the so-called "impossibility" results (see Arrow, 1963; Sen, 1970). Briefly, and roughly speaking:

- If the attempt is made to advise on what is "best" for the society, on the basis of a "general" rule (or set of criteria), then the choice comes down to one between "dictatorship" or "inconsistency."

---

[2] The nonseparability of efficiency and distribution has been discussed at length by several economists; for example, Samuels (1972), Bromley (1990), and Martinez-Alier and O'Connor (1996). A simple mathematical model for the archetypal two-resource, two-good, two-agent general competitive equilibrium economy, where agents have different preferences, is presented by O'Connor and Muir (1995). The tug-of-war between "present" and "future" generations was highlighted by Howarth and Norgaard (1990, 1992); see also Muir (1996) and Faucheux, Muir, and O'Connor (1997).

- If both "dictatorship" and "inconsistency" are to be avoided by weakening the rule system, then either the advice may be indecisive or the possibility is opened of dishonorable outcomes.

There is, nonetheless, a quite simple way of moving forward. This is to reframe the "impossibility" results, taking them as hints of a probable deep structural property of situations of human coexistence and coordination. The suggestion is to recognize that people (including ourselves) are indeed unreconciled, not only to each other but often also within themselves, and that being "shot through with contradictions" is part of being human in society. This does not mean that "anything goes," in the much-misunderstood phrase of Paul Feyerabend (1975). Rather, it follows – as Feyerabend indeed would insist – that there may coexist a plurality (perhaps irreducible) of evaluation or justification principles that, while being all pertinent in some way(s), cannot all be applied simultaneously (or, at best, may lead to divergent recommendations).

In other words, it can be *reasonable not to be rule-bound*. This can be vexatious, but it is not really such a new problem for economists. John Stuart Mill had encountered it many times (see O'Connor, 1995, 1997); environmental philosophers currently discuss it (Stone, 1987; Holland, 1997). Several generations of institutional economists, such as Commons (1934), Kapp (1968, 1969), Schmid (1978), Bromley (1989), and Samuels (1992), have actually insisted on the importance of empirical and theoretical analysis of the instituted processes of "working out" responses to various social choice and coordination dilemmas. For example Commons, in his *Institutional Economics*, taking the cases of legal tribunals, offered an elaborate plea for a process view of economic reasons and reasoning:

> The Court enters beneath the letter of the law and investigates the economic circumstances out of which the conflict of interest arises. Each dispute is a separate case with its own facts, although these facts may be brought within general principles and reconciled with particular precedents discovered in similar cases. The general weighing of all the facts thus investigated, in view of all these principles and precedents, is the process of deciding what is reasonable under all the circumstances. (1934: 712)

Writing in advance of Arrow's mathematical axiomatization, Commons insisted that no "general" formula could be relied upon to produce "reasonable" outcomes in application to all sets of problems of fairness and justice in resource allocation. Reasoned and reasonable compromises would have to be deliberated and worked out in a social process. Moreover, this permanent working out of our coexistence problems centres around the substance and significance given to redistribution of risk and economic opportunities – what Samuels later calls the "distribution of sacrifice" – at any moment in time and projecting into the future.

Fuzzy feasibility stakes

Having sketched the classical problem of social choice, let us return to the ecological and economic systems issues of "feasibility" that delimit and frame the decision problem. Determining what might be feasible in ecological economic futures is partly a matter of science and technological know-how. But uncertainties abound. The "space of feasible outcomes" is characterized *ex ante* by an inherent indeterminacy and *ex post* by irreversibilities. Knowledge in the sense of insight and understanding is not synonymous with capacity for predictions. Awareness of risks is not synonymous with capacity to intervene to reduce or control the risks. Examples currently in the news include: greenhouse gas emissions into the atmosphere and perturbations to climate patterns; cloning processes, where the transmission of cell "biological age" is a complex phenomenon; medical drugs whose "side-effects" are unpredictable in time and from one species to another; genetic splicing and eventual population biology consequences (including the possible cross-fertilization of genetically modified and nonmodified strains of commercial food plants); nuclear fuel cycle experiments; and new chemicals produced, or by-produced, for industrial processes.

Many scientists will argue that ignorance and incompleteness of knowledge have always been admitted within the scientific project. They are partly right. At stake, however, is not the admission of partial ignorance but, rather, the significance to be attached to the forces of change being engaged under conditions of inability to exercise mastery over eventual outcomes (Funtowicz and O'Connor, 1999). In a stylized way, we can observe that the question of society's attitude(s) toward technological progress tends to polarize around a question of the burden of proof:

- Those who evoke the traditional discourses of progress and perfectibility (and others invoking mere adventurism) will argue that "the future can look after itself" and that all risks should be run.
- Those who evoke a "precautionary" attitude will argue about the risk of so-called "Type II Error," emphasizing that absence of proof of danger is not the same as proof of the absence of danger. Where uncertainty and possibly grave dangers reside, the risk should not be run.

In their pure forms, neither of these positions is satisfactory. Often, it is not possible to furnish definitive proof of danger, nor definitive proof of nondanger. Some risks must be run (otherwise there are the dangers and contradictions of paralysis, and so on). Yet, a heedless rush into ecological, geophysical, metabolic, and chemical novelty seems (to many people) an excessive enthusiasm for making trouble. So, we have an interesting – some would say impossible – situation in which, strictly speaking, neither rule can be applied; yet each precept acts as a caution on (or, indeed, a refutation of) the other, creating a sort of dilemma or impossibility. This is the hallmark of environmental governance problems. It is, in other words, impossible to go beyond this sort of situation of contradictory imperatives, or contradictory counsels of "good reasons."

But this does not mean that a "reasoned" base for policy is impossible. Rather, if reasoned basis for action is to be established, then forms of deliberative and regulatory procedure must be established, that "relativize" the contradictory positions while not seeking entirely to dispose of any of them. The challenge would be to work with a permanent "argumentation" between the two – or more – contradictory positions. In such circumstances, an analyst needs to be like a "mid-wife of problems" (Rittel, 1982: 35–48), helping to raise into visibility, "questions and issues towards which you can assume different positions, and with the evidence gathered and arguments built for and against these different positions."

Take the cases of toxic products, long-lasting active wastes, or novel products having a permanent perturbation potential (such as genetic recombination products). We may propose the application of the following rule placed as a requirement upon any scientist or decision-maker or innovator promoting a new product or waste management solution: "Consider the possible significance of 'Type II error' and justify publicly your decision to neglect it."

The sense of this rule is deliberately paradoxical, seeking to bring both imperatives into confrontation with each other, creating something of a double-bind. It is well known that double-binds can create nervousness and moroseness. Yet, they can also be the contexts of courageous and principled action. This is important, because the environmental dangers in question can implicate large populations and diverse interests across society.

Moral justifications for deliberation

The typical sustainability "social choice" problem – characterized by distributional conflicts and uncertainty – appears to lead to a bifurcation point, at which a person or group will be required to choose between two forms of discourse and action:

- On the one hand, discourses (seeking to be translated into practices) of *domination*, corresponding to Arrow's notion of "dictatorship." This means the exclusion or discounting of any contradictory principles of what is good and should be done, a purely strategic concern (in order better to dominate) any evidence of "other points of view."
- On the other hand, discourses taking up a challenge of tolerance – proposing to search out possibilities of *coexistence* based on respectful consideration of a plurality of antagonistic or seemingly contradictory considerations. As the Buddhists say, "Do not take life unnecessarily."

The first option has a tendency to simplify toward "Might is right," but also to support a variety of discourses of unreflective moral zeal on the part of self-identifying elites (and also, indeed, of some persecuted minorities).

The second option is more complex. Its realization, in any real situation, is at best an open question, since it wagers on the possibility of some sort of mutual respect that acknowledges a real dissent between various contending interests and principles of justice or justification, and yet the possibility of honorably living/working within this dissent. Even if everybody affirms such a commitment to seek out possibilities of accommodation, compromise, coexistence, there is no certainty of avoiding outbreaks of hate, despair, war, impatience, and intolerant violence.

One might look for the possibilities of dialog, reciprocal learning, accommodations and adaptation, and discursive and deliberative processes for making visible the diversity and seeking a reciprocal awareness – even a reciprocal evaluation! – of this plurality of "reasonable" claims and (sometimes incompatible) points of view. However, this merely puts on stage the problem of coexistence; it may highlight the tensions and the contradictions, but it does not in itself put an end to them (Latouche, 1984, 1989; O'Connor, 1999a, 2000). We would be confronted with the eternal fragility of this coexistence even if it came to be widely affirmed as an "ideal."[3]

At the risk of simplifying rather a lot, it is useful to explore further this contrast of a *domination ethic* as compared with a *coexistence ethic* (Salleh, 1990, 1997; O'Connor, 1999a). A domination ethic tends to consider the outside world, including other people, as means to an end, and/or as obstacles to achieving one's purposes. A coexistence ethic, by comparison, would seek out forms of courtesy and dialog. There is tolerance of tensions, admission of antagonisms as legitimate but to be dealt with on the basis of a desire for a coexistence. Yet, an attitude of courtesy, hospitality, and welcoming in a spirit of coexistence cannot be taken for granted, nor does it spell out a magical harmonization. In this regard, as Latouche suggests, the conviction in the merits of a philosophy of coexistence can arise almost paradoxically:

> . . . as there is no hope of founding anything durable on the short-change of a pseudo-universality imposed by violence and perpetuated by the negation of the other party, the venture is warranted that there is indeed a common space of fraternal coexistence yet to discover and construct. (1989: 139)

Revealing the Social Demand for Reconciliation

The Prisoner's Dilemma and the social demand for sustainability

Why should I show tolerance, or courtesy, toward my neighbor? Why should we strive for coexistence rather than conquest and eradication (as we prefer for

---

[3]   Some readers will discern a parallel with themes of Jürgen Habermas, notably communicative rationality, emancipative discourse, and deliberative processes characterized by the "ideal speech situation" (Habermas, 1979). What our perspective shares with Habermas, and with those who have followed his themes, is interest in the proposition that, starting from a situation of conflicts, dissent, misunderstandings, and antagonism, some reconciliation is possible through a process of dialog and deliberation. Habermas also recognizes that a willingness to listen and reflect cannot be taken for granted – it involves a "choice" of ethical stance, with an existential and cultural contingency. On the other hand, whereas Habermas would (it sometimes seems) like to hope that the reconciliation will emerge through uncoerced force of reasoned argument, we are sure that there is no guarantee that reason will lead to reconciliation. We place our hopes, just as much, on affective as well as reasoned motivations in *the search for* a coexistence: that is, the constitutive role played – for better and for worse – by the human sentiments or passions. Having lived through the Holocaust, Habermas no doubt harbors some reserves as to where populism, sentimentalism, and nationalistic enthusiasms might end up. Yet, the mobilizing and constitutive force of sentiments should not be neglected (partly because, if neglected, they are rather likely to reemerge in undisciplined ways).

malarial mosquitoes)? Why should I care for the fortunes, or misfortunes, of future generations of neighbors? Even if I love my neighbor, what is to stop the other neighbors from cheating on him or her and me (and on future generations)? Our common-property environmental problems have some difficult starting-points. The basis for acting/choosing in these situations is not self-evident – or, at least, what seems self-evident to one person is not agreed to be so by others . . .

We may present this as a kind of Prisoner's Dilemma. For example, information in many parts of the world suggests that many water resources are becoming contaminated, sometimes irreversibly, with pesticides, industrial wastes, and other toxic chemicals. Scientific analyses suggest that, as a result of enhanced greenhouse gas emissions, the atmosphere is warming up. But, why should people – as citizens, as consumers, as farmers – be concerned about the degradation of ground water quality, or (even less) about climate change? Why worry about greenhouse gas emissions (yours, mine, or anyone else's)? As Groucho Marx once might have said, "What have future reduced GHG emissions ever done for me?" Why should I choose an action that cooperates with the future, since I cannot expect the future to cooperate with me?

This leads on to the question: What are the prospects for social learning, participatory and deliberative procedures for decision support, policy definition, and evaluation, such that people may be encouraged, one and all, to "jump out of the Prisoners' Dilemma" (Guimarães Pereira and O'Connor, 1999)? If sustainability is to become a guiding reality, new forms of politics will have to be invented that seek out prospects of coexistence. There is currently much debate over the extent to which democratic political process can or should allow for reflective deliberation, and how this might be reflected in pursuit of sustainability (Dryzek, 1994; Holland, 1997; Sagoff, 1998). The variety of participative and deliberative procedures is very wide: however, one widespread feature is the emphasis on tolerance and coexistence of divergences. At the heart of any notion of democracy or deliberation is the admission of a plurality of potentially "reasonable" views and claims on the situation, which should be listened to, before rushing to a decision.

To take this theme further, we will take the topical question of the potentials of the new digital information and communication technologies (ICT) as tools for deliberation and decision support in the environmental policy domain. After a brief discussion of the information requirements for framing sustainability policies (second and third subsections), the example of aquifer water resources will be developed to illustrate prospects for "revealing the social demand for reconciliation" (fourth and fifth subsections). The purpose is to highlight how it is possible to work in a deliberative way – that is, through procedures of stakeholder concertation – with the feasibility (systems potential) and desirability (social choice) questions that are at the heart of sustainable development.

Information requirements for resolving sustainability problems

Principle 10 of the UNCED Declaration, made at Rio de Janeiro in 1992, affirmed that "environmental issues are best handled with the participation of all

SYSTEMS SCIENCE (feasibility)	Information, indicators (and uncertainties)	(virtual) ICT INTERFACES	Social actors (stakeholders)	SOCIAL SIGNIFICANCE (desirability)
Knowledge, resources, and techniques	Analytic methods for option appraisal	(real) POLICY ISSUES (Sustainability of what and for whom?)	Motivations, interests, and justifications	Ethics, culture, and values

Figure 12.2 Complementarity of investigations of systems potentials and social desirability.

concerned citizens, at the relevant level." Yet, learning about economic and environmental issues involves confrontation with a diversity of objectives and interests that are expressed in a variety of vocabularies and at different scales. Information and communication frames must be developed not just with a view to scientific validity, but also from the standpoint of the ways in which they help (or don't help) to "set the stage" for convivial exchanges of perspectives. What is most critical is appreciation of the *significance* to different groups and persons of alternative resource management choices (or even, in some cases, choices to *not* manage particular processes, ecosystems, and resources).

Figure 12.2 highlights the complementarity between, on the one hand, investigations of systems potentials or feasibility and, on the other, investigations of the criteria of desirability or social choice for feasible courses of action.

The *systems potential* aspect can be seen as a generalization of the traditional economics question of supply costs, adapted to the long-timescale and larger-system perspectives that characterize sustainability concerns. Economic resource management must fulfill two complementary functions. The first is the delivery of economic welfare in the narrow sense, through production of economic goods and services; the second is the maintenance of the ecological welfare base through assuring reproduction or enhancement of critical environmental functions.[4] Policies aimed at safeguarding the support functions of the environment require the commitment (or reorientation) of scarce resources. Sustainability objectives can thus be thought of as responding to a kind of social demand for the maintenance of environmental functions. As discussed in Part I, this social demand for environmental quality and for assuring fairness toward future generations (including protection from future harms), cannot easily be reduced to simple monetary values. Rather, scenarios that explore different conceivable co-evolutions of ecological and economic systems need to be formulated and evaluated from various

---

[4] Environmental functions are here defined as any capacity or performance of natural processes that assures the permanence of living ecosystems and/or furnishes goods and services of value to human society. The justifications for produced economic output and the maintenance of environmental functions as complementary sustainability criteria – the so-called "strong sustainability" perspective – are developed by, among others, Hueting (1980), Faucheux and O'Connor (1998), and Ekins and Simon (1999).

points of view. These include scientific preoccupations such as sensitivities to speculative hypotheses about technological capacities and ecological systems changes, and also societal preoccupations that can be summarized in the phrase "Sustainability of what, and for whom?" These analyses will usually entail various forms of systems representation, simulation modeling, and quantification that integrate economic and ecological components, notably:

- *statistically aggregated economic information* – such as systems of accounts and models that quantify volumes of sectoral production, water use, and greenhouse gas emissions on a national, regional, or world basis
- *spatially defined environmental information* – such as an aquifer or watershed, or the global atmosphere considered as a fluid dynamic circulation system, coupled to the oceans, which is being "forced" by the inflow of anthropogenic greenhouse gases

This is the realm now known as integrated modeling, which combines ecological and economic dimensions, and has now become a major activity of interdisciplinary policy-relevant research endeavor.

The *social choice* problem is to decide what might be desirable within the bounds of the feasible. Abstractly, this takes on the form of an arbitrage between different interests, just as in Arrow's classic formulation. Following the Brundtland formulation (World Commission on Environment and Development, 1987), we can consider the specific problematic of sustainability as a tension between present and future generations. And, in the context of environmental valuation problematics, this in turn can be seen as one aspect of a more generalized structural opposition – between "us" and the "others," between self-interest and interest in the livelihoods of others, between human and nonhuman communities, between "our" culture (whichever it is) and other cultures, and so on (see Arnoux, Dawson, and O'Connor, 1993; O'Connor, 1994, 1999b; Salleh, 1997; Hailwood, 2000). The variety of candidate sustainability ethics that, over the years, have been put forward, tend indeed to turn around this time-honored problem of reconciling concern for oneself with a consideration for the other(s). This suggests that two forms of social information or representation will have special pertinence for a deliberative approach to resource valuation and governance:

- *local-level information* – that is, the immediate life experience of "ordinary" members of society, in their homes, workplaces, farms, shops, schools, with friends, and on their travels
- *governance information* – the terms in which a regulation and coordination of human action is conceived, that link local and aggregated economic and ecological information to frameworks of collective purpose, responsibilities, conflict management, and policy implementation[5]

---

[5] It has become commonplace to refer to economic, ecological, and social dimensions of sustainability. The "social" dimension has often remained rather amorphous, and often drifts back toward

The above formulation thus distinguishes four basic dimensions of information: ecological and economic systems information, individual knowledge, and governance or institutional framing information. These may be considered as irreducible dimensions for building a good representation of an environmental issue. (There are, of course, many, many "local" standpoints.) The challenge is to find ways of representing the systems feasibility information in a way that orients individuals toward an awareness of the higher-level institutional process of resource governance. This requirement can be summarized in the following formula: a good indicator set must signal (or reinforce, etc.) the existence or creation of plausible and convincing institutional arrangements for coordinating the actions of all involved parties in a fair and acceptable solution for the pursuit of the sustainability goals.

We may specify two qualities that need to be satisfied by any representation or category of information (such as a number produced in a valuation study or an image on a video screen) if it is to perform effectively in the desired role of supporting stakeholder deliberation on sustainability problems:

- First, the indicator, image, or whatever should mark a passage between different scales of representation of an economic, ecological, or political coordination situation (from an individual to a more aggregated perspective).
- Second, the indicator, image, or whatever ought to speak meaningfully to at least two (or more) different categories of stakeholders; namely, it ought to find a meaningful place within a plurality of distinct "life-worlds" or decision contexts.

These are necessary; they may not be sufficient conditions. Prospects for framing and promoting sustainability policy choices as collective and concerted actions can, we suggest, be enhanced through bringing the different scales of information and different stakeholder perspectives into constructive confrontation with each other. At a scientific level, this means establishing "bridges" between representations at different levels of aggregation or based on varied conceptual frameworks. At the social level, it means building the capacities for mutual understanding of the contrasting perspectives and preoccupations of different stakeholders, in order to search for points of common ground.

economic information, such as employment, income, and property ownership. More recently, though, emphasis has been placed on the political/institutional dimensions. For example, the FAO (1999), in work on indicator systems for sustainable management of fisheries, designates economic, social (local), ecological, and governance dimensions, thus drawing attention to the institutional basis for resolving the problems of social choice. A "tetrahedral" framework for integrated representation of systems potential and social choice problems is being applied by a number of European research groups for the development of ICT, notably in the domains of climate policy (Guimarães Pereira and O'Connor, 1999), water and soil pollution from agriculture (Douguet, O'Connor, and Girardin, 1999; Douguet and Schembri, 2000), and marine fisheries, forests, and underground water resources (work in progress at the C3ED).

Deliberation in the "theater of sustainability"

Take, for example, the governance challenges of freshwater resource exploitation and conservation. The fundamental distribution questions are easy to articulate. In the absence of acceptance that future generations might have an "entitlement" to high-quality water resources and to wetlands as a heritage value, it may appear socially optimal (that is, Pareto-efficient, with the rights distribution skewed toward the present generation of users) to deplete or irreversibly degrade some water resources.

In a general way, policies for water system management (ecological conservation, irrigation, urban supply, industrial use, river water flow control, draining and building, sewage disposal and pollution monitoring and control, and so on) will involve choices for the redistribution through time of economic opportunity and of access to services and benefits provided by the biophysical environment. Water cycles and flow patterns, including the underground zones and transportation, are also part of the ecosystem infrastructures that support habitats of mountainside, swamp, riverbank, and aquatic species. The water may be a potentially valuable input for industrial, agricultural, and urban consumption. But if aquifer reserves are exploited, or river water flow is diverted for irrigation, for factory use, for power plant cooling, or for urban drinking supply purposes (for example) – or if the continuity of flow is interrupted through dams, reservoirs, and other forms of storage – the natural forms of life may be put at risk. Water that has been used for economic purposes may be allowed to flow back into natural systems in a dirtied or polluted condition; this also can menace the viability of life forms and can pose problems for human health.

The application of principles of stewardship, precaution, and fairness in distribution may be explored in a general framework of scenario or "futures" studies. Tensions, conflicts of interests, uncertainties, and dissent amongst scientists, as well as governance challenges, can be explored by cross-comparison of different scenarios about regimes of water resource use and corresponding institutional arrangements:

- One set of scenarios would usually be trend-based or "business-as-usual" projections, which may often involve trends in water use that are unsustainable.
- Other scenarios may then be constructed that involve the satisfaction of specific sustainable use criteria, on the basis of various hypotheses about systems potentials and about social choices of "what, and for whom?"

In order to preserve a conceptual link to the established forms of benefit–cost analysis (BCA), it is useful to note that this new style of scenario-based evaluation is an extension, to new terrain, of the well-established fundamentals of welfare economics concerning the inseparability of allocative (efficiency) and distributional (equity) goals. The extension takes into account two key points, as follows. (1) The further that concerns of environmental policy extend into the long-term future, the more will intertemporal distributional considerations predominate over allocative efficiency in policy formulation and appraisal. (2) The further that concerns for environmental values extend into the domains of aesthetic and cultural, as well as economic, appreciation of natural cycles and systems, the more difficult it becomes to obtain meaningful monetary valuation estimates based on the assumptions of value-commensurability and substitutability that underlie the established BCA approaches. With this backdrop, the structure of our approach to valuing water resource uses may now be represented in the following terms:

- First, the normative dimension inherent in the sustainability referent is reflected in the way in which scenarios are formulated explicitly as respectful, or not, of fundamental notions of social, economic, and ecological sustainability.
- Second, substantive attention is given to inter-group and intragenerational distribution issues by the requirement to give a specific content to the sustainability goal, through description of, and analysis of, possible incompatibilities between the diverse sustainability concerns expressed by the variety of stakeholders (cf., O'Connor and Martinez-Alier, 1998).
- Third, opportunity costs for alternative water uses can be estimated with reference to any specific scenario for water uses. This corresponds, roughly speaking, to the partial equilibrium type of analysis for small variations around a presumed economic structure.[6]

Having established the general conceptual orientation, the next task is to specify an institutional and deliberative context. Information about interests and priorities can be built and debated in a "theater of sustainability" – as suggested schematically by figure 12.3. A stakeholder concertation process can be developed that integrates systems science with deliberation in a recursive cycle as follows. The portrayal of an iterative loop is intended to emphasize the real-time process of *putting on to the scene* interests, knowledge, disagreements, and possible solutions. The first step in the cycle privileges the *social choice* (or desirability) preoccupations at the stakeholder level; the next four steps privilege the *systems potential* (or feasibility) aspect of analysis; and the last two steps again privilege the *social choice* problem, this time at the governa. ce level.[7]

By starting with the social significance axis of learning (Step 1), it is emphasized that the information and appraisal requirements for water resource/

---

[6] In cases in which, using comparable data bases and estimation methods, the relative valuations of different uses are similar across all scenarios, it can be said that the valuations are fairly insensitive to distributional considerations. Conversely, where the relative valuation changes significantly as a function of the scenario adopted, it is revealed that distributional variables (and, behind these, possible differences of principle between groups of stakeholders) are the most significant ones for water use choices at all levels.

[7] An example of Steps 3 and 4, the appraisal of environmental functions associated with freshwater resources, in the region of Bretagne in western France, is given by Douguet and Schembri (2000). An example of scenario-based multicriteria appraisal incorporating stakeholder concertation, for water resource futures in the district of Troina, Sicily, is given by De Marchi et al. (2000).

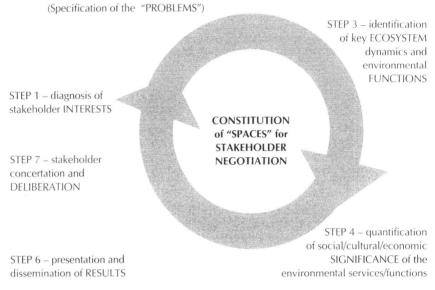

STEP 2 – SCIENTIFIC ANALYSES of the hydrosystem(s) and ecosystems (e.g., river/wetland/marine ecology, land/water quality, biodiversity, chemical contamination, microbiology, etc.)

(Specification of the "PROBLEMS")

STEP 3 – identification of key ECOSYSTEM dynamics and environmental FUNCTIONS

STEP 1 – diagnosis of stakeholder INTERESTS

CONSTITUTION of "SPACES" for STAKEHOLDER NEGOTIATION

STEP 7 – stakeholder concertation and DELIBERATION

STEP 4 – quantification of social/cultural/economic SIGNIFICANCE of the environmental services/functions

STEP 6 – presentation and dissemination of RESULTS

STEP 5 – SOCIOECONOMIC ANALYSIS of resource management and development OPTIONS (evaluations via multicriteria and forward studies (scenario) analyses)

Figure 12.3 The deliberation cycle in the "theater of sustainability."

environmental governance are grounded in specific contexts of learning and action. These will include both formalized and "informal" knowledge, the latter being typically held by members of local networks and communities (including retailers, financial and agrobusiness services, and so on) without necessarily being abstracted or theorized into systematic models. Interactive stakeholder-linked approaches imply the need to present and discuss scientific and socio-economic findings to interest groups with a range of different interests, on a permanent (recursive) basis. It is here that the new interactive ICT can be particularly effective.

Exploiting ICT for framing social choices over water resources

Researchers and teachers all over the world are currently exploring the use of information and communication technologies (ICT) as a medium for organizing economic and environmental information so as to respond to qualitatively different educational, analytic, and normative circumstances. Multimedia ICT products typically permit individual use (such as from a CD-ROM or via web-site access). They also imply that the user is a member of a larger community. Learning is always a social process with its many contexts (geographical, institutional, and so on), and individuals participate in collectivities through various forms of intersubjective communication. A convivial or user-friendly ICT video interface links a person within his or her "own place" to other spaces of life, other forms of information, interests, interest groups, and policy analyses, via corridors of translation and reciprocal appreciation.

Two forms of computer-based representation can be considered as the key stage props that help to bring water governance problems on to the stage in the theater of sustainability. Using the terminology introduced by Guimarães Pereira and O'Connor (1999), these are: (i) *personal barometers*, which allow quantification of environmental impacts of individual lifestyles; and (ii) *scenario generators*, which allow individual, firm, or household unit activities to be put in the context of possible future trends and changes in patterns of economic activity and in the state of the environmental resources. Taken together, the two will consist of a family of models and visual representations that allow the quantification of environmental impacts linked (directly or indirectly) to personal consumption and lifestyle, and also the specification of scenarios developing different perspectives of "what is sustainable" in economic and environmental terms. The governance challenges can be brought into focus by this process of visualization.

We illustrate the general idea with reference to exploitation of common-pool aquifer resources.[8] According to our design concepts, the primary user interfaces should be developed at the "local" level of knowledge categories; for example, individual water use by farmers or households may be quantified. The purpose of a personal barometer will be to allow people, interacting with the computer, to specify their personal contributions (direct and indirect) to aquifer exploitation, and to begin to reflect on the wider context of these individual actions in terms of both systems and social significance.

Imagine the sorts of responses and conversations that might eventuate when a farm owner, manager, or worker responds to the question, "How much water do you use?" In the "conversation" mediated by the ICT, various types of answers might be forthcoming, and this will raise questions about the factual reliability of the information supplied and the purposes of supplying the information. One can invent a simple classification, namely (1) "Don't know," (2) "Deliberate dissimulation," and (3) "Telling the truth."

If scientists are seeking data to calibrate their models, this ambiguity in responses is a real headache. However, if emphasis is placed on understanding the resource problem in terms of people's interests, motives, and social relations, the complexity of the communication situation can be viewed positively. Table 12.1 invents possible reasons (for illustrative purposes), being combinations of circumstances and motives that might correspond to each category of response. This illustrates that all communications of "information" will, in a variety of ways, be grounded in economic interests and social relations. Economists will often emphasize strategic considerations about the release or withholding of information, and about the falsification of information. Anthropologists might

---

[8]   There is a huge body of literature on water resources politics (see McCully, 1996). For two good examples on underground resources, see Aguilera-Klink, Perez Moriana, and Sanchez Garcia (2000) and Allal and O'Connor, in Lonergan (1999).

Table 12.1 Information and dis-information – the play of reasons and interests

Response to the question "How much water do you use?"	Transparent meaning, possible explanations	Dissimulation, possible explanations
"I don't know"	The respondent really doesn't know (there is probably no water metering or charging system)	The respondent has a fairly good idea, but doesn't want to say (for example, he or she may fear use of the information against his or her interests, or may believe that he or she should be paid to supply information)
Much higher than the real level	A sincere response based on miscalculation or misperception	A strategic response that constitutes a claim on the water (for example, a water regulation scheme is being contemplated where property rights may be based on historical usage)
Much lower than is really the case	A sincere response based on miscalculation or misperception	A strategic response that constitutes a social positioning statement. For example, the respondent may want to convey the image of an efficient or frugal use of water (in order to avoid shame, perhaps)
Close to the real situation	The respondent knows and tells (for example, operates pumping technologies that permit monitoring)	The respondent is really unsure, but doesn't want to admit this, and just by chance has estimated a figure that is close to reality

put more emphasis on the negotiation by individuals of their social relations and status within communities and networks.

This example shows information not as a quantity but, rather, as a social process of building and negotiating meanings and capacities of action. In a deliberative approach to the construction and resolution of resource management problems, one may exploit the strategic and social relations factors in order to "build a common problem." ICT interface capacities can be exploited so as to situate water users in relation to the collective problem. The answers that individual water users give, and the reasoning underlying their responses (and their nonresponses, their silences, and their dissimulation), all convey something about the socioeconomic realities and the stakes.

With the use of ICT, branching out from a personal barometer, it is possible to set individual water use figures visually (by images, graphs, or commentaries) in comparison with figures for local or regional averages. These figures – such as amounts of irrigation water for maize in western France – can be presented in comparison to figures for other crop types (such as wheat) and for other countries

(Israel, for example, where water-economizing technologies are very advanced). Respondents can also compose their own estimates. For example, an individual farmer's figures can be multiplied up, on the basis of data (or guesses) about the numbers of users in each category, in order to obtain a figure for aquifer exploitation at the whole catchment level. Using the Internet, or printed sources, other information – or guesses – can be brought in, about the full range of water exploitations (including extraction for industrial use, if any, and for town water supplies, and so on) and about the aquifer's recharge or carrying capacities. In this way, a process of reflection is created in which the individual farm activity is placed within the greater economic and hydrological scheme of things.

Moving forward from these initial phases of building and reconciling pictures of the present situation, explorations can be made of possible – and perhaps desirable – futures. This is where the scenario generator concept comes in. Suppose that, according to individual farmers, or fertilizer companies, or the Chamber of Agriculture, production of maize is expected to increase by a factor of ten over the next 15 years. What will this imply for water demand? The water consequences of economic development scenarios can then be set in confrontation with observations and hypotheses gleaned from hydrosystem modelers and farmers' own observations about aquifer storage volumes, recharge and replenishment rates, water table changes, and so on.

If this process takes place interactively, then individuals' water use figures, guesses, estimates, and so on are exposed to reciprocal scrutiny. This brings about the possibility, for each ICT user, of assessing others' assumptions and of evaluating information claims. People may interrogate each other, in more or less convivial fashion ("I don't believe it," "That's not possible!" "You'll need more water than that . . . ," and so on). The ICT users may be led to reflect on, and debate with others, the assumptions made about individual uses – including their own, present and future – and about sectoral developments, aquifer recharge rates, water table, and wider ecological consequences (riverside vegetation, fish populations in rivers, and so on). They are led to identify impossibilities and contradictions. These may include seeming systems impossibilities (how to really extract 5 million m^3 of water annually from an aquifer whose recharge rate is estimated at 2 million m^3 ± 50 percent for an average year . . .). There may also be social and economic impossibilities, such as where to sell all the maize at a worthwhile price.

Through this sort of visualization and futures exploration process, people are confronted with the *systems potential* and *social choice* aspects of their common problems. Farmers, water users, householders, or – as representatives of sectoral interest groups – local political leaders, chiefs of water companies, catchment authorities, environment ministers, and NGOs, must confront the question of managing conflicts within the limits of what is feasible. Inasmuch as the ICT can facilitate learning and a sharing of perspective, the process of problem representation can be the point of departure for a deliberative search for sustainable use solutions based on restraint, respect of divergent criteria, and the acceptance of a principle of coexistence. In this, they could follow the reasoning of John Stuart Mill who, a century ago, affirmed that:

> When the "sacredness of property" is talked of, it should always be remembered, that any such sacredness does not belong in the same degree to landed property. No man made the land. It is the original inheritance of the whole species. When private property in land is not expedient, it is unjust. It is no hardship to any one to be excluded from what others have produced: they were not bound to produce it for his use, and he loses nothing by not sharing in what otherwise would not have existed at all. But it is some hardship to be born into the world and to find all nature's gifts previously engrossed, and no place left for the newcomer . . . (1909: 230)

Concluding Remarks

Environmental resource management is, *par excellence*, the domain of "common problems" – that is, situations of strong and visible interdependence between individual and collective actions, characterized by the Prisoner's Dilemma. The resolution of such problems means dealing incessantly with moral choices, and this makes calculation, measurement, and technical expertise on their own insufficient. Decision quality assurance and socially legitimate governance processes can be assured only through integrating scientific, technical, and economic expertise within a permanent stakeholder communication process, in order to search for common ground.

However, there will evidently be many situations in which people, or different cultures, or different species of plants and animals, simply cannot, or do not want to, find a basis for durable coexistence. Therefore, reflective deliberation, as advocated here, may work to highlight appreciation of tensions, but it does not necessarily find a way to put an end to them.

If one party does not want to seek out some form of coexistence, it may be because it holds an ethic of exclusion, or of "domination." Or it may be because – in the specific circumstances that present themselves – it interprets other parties as a mortal threat that has to be resisted, notwithstanding a general disposition toward "coexistence." (Malaria and human populations have an uneasy coexistence.) Or it may be that the differing experiences of the coexisting parties are incomparable, being grounded in different existential conditions and – in the case of self-conscious beings – in different ethical and epistemological postulates that, each in their own terms, are reasonable.

It is worthwhile reflecting that some of the almost-silenced nondominant voices and populations actually express a real commitment to tolerance under the most adverse conditions, and thereby put to shame many more powerful purveyors of rationality. The various local subsistence societies that still exist around the world have a lot of trouble coexisting with modern mass-consumption society, which is based on globalized oil, mineral mining, forest cutting, agrochemical, and fisheries depletion activities. Up against the wall of the World Trade Organization, local communities and subsistence peoples might decide to revolt (which, on the basis of past experience, has rarely brought durable success). Or they may set out to see what new opportunities the globalization adventure can offer to them. Or they may well foresee their fate as being wiped out, with what remains of their dignity. These are all ethically and culturally coherent choices; the point is simply made that a coexistence ethic does not necessarily make for easy living. The "coexistence" ideal of a dignified compromise does not mean finding, by some magical process of option creation, a win–win outcome in which everyone takes away from the negotiating table a large part of what they came to bargain for. Rather, it means reciprocal consideration, the acceptance of sacrifices in a spirit of coexistence, and the ability to refine and change one's personal (or group, or national) goals in the interests of the wider community.

References

Aguilera Klink, F., Perez Moriana, E., and Sanchez Garcia, J. 2000: The social construction of scarcity: the case of water in Tenerife (Canary Islands). *Ecological Economics,* 34, 233–46.

Allal, S. and O'Connor, M. 1999: Water resource distribution and security in the Jordan–Israel–Palestinian peace process. In S. C. Lonergan (ed.), *Environmental Change, Adaptation, and Security.* Kluwer: Dordrecht, 109–29.

Arnoux, R., Dawson, R., and O'Connor, M. 1993: The logics of death and sacrifice in the resource management law reforms of Aotearoa/New Zealand. *Journal of Economic Issues,* 27, 1059–96.

Arrow, K. 1963: *Individual Values and Social Choice,* 2nd edn. New York: John Wiley.

Bromley, D. W. 1989: *Economic Interests and Institutions: The Conceptual Foundations of Public Policy.* Oxford: Blackwell.

Bromley, D. W. 1990: The ideology of efficiency. *Journal of Environmental Economics and Management,* 19(1), 86–107.

Commons, J. R. 1934: *Institutional Economics: Its Place in Political Economy.* Madison, WI: University of Wisconsin Press (reprinted in 1961).

De Marchi, B., Funtowicz, S., Lo Cascio, S., and Munda, G. 2000: Combining participative and institutional approaches with multicriteria evaluation. an empirical study for water issues in Troina, Sicily. *Ecological Economics,* 34, 267–82.

Douguet, J.-M. and Schembri, P. 2000: *CNC: Quantification et Modélisation du Capital Naturel Critique pour la mise en œuvre d'une politique du développement durable en France.* Guyancourt, France: Université de Versailles St-Quentin-en-Yvelines, C3ED Rapport de Recherche.

Douguet, J.-M., O'Connor, M., and Girardin, P. 1999: *Validation socio-économique des indicateurs agro-environnementaux.* Guyancourt, France: Université de Versailles St-Quentin-en-Yvelines, C3ED Rapport de Recherche.

Dryzek, J. 1994: Ecology and discursive democracy: beyond liberal capitalism and the administrative state. In M. O'Connor (ed.), *Is Capitalism Sustainable? Political Economy and the Politics of Ecology.* New York: The Guilford Press, 176–97.

Ekins, P. and Simon, S. 1999: The sustainability gap: a practical indicator of sustainability in the framework of the national accounts. *International Journal of Sustainable Development,* 2(1), 32–58.

Faucheux, S. and O'Connor, M. (eds.) 1998: *Valuation for Sustainable Development: Methods and Policy Indicators.* Cheltenham, UK: Edward Elgar.

Faucheux, S., Muir, E., and O'Connor, M. 1997: Neoclassical theory of natural capital and "weak" indicators for sustainability. *Land Economics,* 73, 528–52.

Feyerabend, P. 1975: *Against Method.* London: New Left Books.

Funtowicz, S. O. and O'Connor, M. 1999: The passage from entropy to thermodynamic indeterminacy: a social and science epistemology for sustainability. In K. Mayumi and

J. Gowdy (eds.), *Bioeconomics and Sustainability: Essays in Honour of Nicholas Georgescu-Roegen*. Cheltenham, UK: Edward Elgar, 257–86.

Guimarães Pereira, A. and O'Connor, M. 1999: Information and communication technology and the popular appropriation of sustainability problems. *International Journal of Sustainable Development*, 2(3), 411–24.

Habermas, J. 1979: *Communication and the Evolution of Society*. London: Heinemann.

Hailwood, S. 2000: The value of nature's otherness. *Environmental Values*, 9(3), 353–72.

Holland, A. 1997: The foundations of environmental decision-making. *International Journal of Environment and Pollution*, 7, 483–96.

Howarth, R. and Norgaard, R. 1990: Intergenerational resource rights, efficiency, and social optimality. *Land Economics*, 66, 1–11.

Howarth, R. and Norgaard, R. 1992: Environmental valuation under sustainable development. *American Economic Review Papers and Proceedings*, 80, 473–7.

Hueting, R. 1980: *New Scarcity and Economic Growth: More Welfare Through Less Production?* Amsterdam: North-Holland.

Kapp, K. W. 1968: In defense of institutional economics. *Swedish Journal of Economics*, 70, 1–18. Reprinted in K. W. Kapp 1974: *Environmental Policies and Development Planning in Contemporary China and Other Essays*. The Hague/Paris: Mouton/Maison des Sciences de l'Homme, 149–67.

Kapp, K. W. 1969: On the nature and significance of social costs. *Kyklos*, 22, 334–47.

Latouche, S. 1984: *Le Procès de la Science Sociale*. Paris: Anthropos.

Latouche, S. 1989: *L'Occidentalisation du Monde: essai sur la signification, la portée et les limites de l'uniformisation planétaire*. Paris: La Découverte. English translation: *The Westernisation of the World*. London: Polity Press.

Lonergan, S. C. (ed.) 1999: *Environmental Change, Adaptation, and Security*. Dordrecht: Kluwer.

Martinez-Alier, J. and O'Connor, M. 1996: Distributional issues in ecological economics. In R. Costanza, O. Segura, and J. Martinez-Alier (eds.), *Getting Down to Earth: Practical Applications of Ecological Economics*. Washington, DC: Island Press, 153–84.

McCully, P. 1996: *Silenced Rivers: The Ecology and Politics of Large Dams*. London: Zed Books.

Mill, J. S. 1909 [1871]: *Principles of Political Economy, with Some of Their Applications to Social Philosophy*, 7th edn. London: Longman.

Muir, E. 1996: Intra-generational wealth distributional effects on global warming cost benefit analysis. *Journal of Income Distribution*, 6(2), 193–214.

O'Connor, M. 1994: Valuing fish in Aotearoa: the treaty, the market and the intrinsic value of the trout. *Environmental Values*, 3, 245–65.

O'Connor, M. 1995: La réciprocité introuvable: l'utilitarisme de John Stuart Mill et la recherche d'une éthique pour la soutenabilité. *Economie Appliquée*, 48(2), 271–304.

O'Connor, M. 1997: J. S. Mill's utilitarianism and the social ethics of sustainable development. *The European Journal of the History of Economic Thought*, 4(3), 478–506.

O'Connor, M. 1999a: Dialogue and debate in a post-normal practice of science: a reflection. *Futures*, 31, 671–87.

O'Connor, M. 1999b: Mana, magic and (post-)modernity: dissenting futures in Aotearoa. *Futures*, 31, 171–90.

O'Connor, M. 2000: Pathways for environmental valuation: a walk in the (Hanging) Gardens of Babylon. *Ecological Economics*, 34, 175–94.

O'Connor, M. and Martinez-Alier, J. 1998: Ecological distribution and distributed sustainability. In S. Faucheux, M. O'Connor, and J. van der Straaten (eds.), *Sustainable Development: Concepts, Rationalities and Strategies*. Dordrecht: Kluwer, 33–56.

O'Connor, M. and Muir, E. 1995: Endowment effects in competitive general equilibrium: a primer for policy analysts. *Journal of Income Distribution*, 5, 145–75.

Rittel, H. 1982: Systems analysis of the "first and second generations." In P. Laconte, J. Gibson, and A. Rapoport (eds.), *Human and Energy Factors in Urban Planning*. The Hague: Martinus Nijhoff, 153–84.

Sagoff, M. 1998: Aggregation and deliberation in valuing environmental goods: a look beyond contingent pricing. *Ecological Economics*, 24, 193–213.

Salleh, A. 1990: Living with nature: reciprocity or control? In J. R. Engel and J. G. Engel (eds.), *Ethics of Environment and Development*. London: Belhaven Press, 245–53.

Salleh, A. 1997: *Feminism as Politics: Nature, Marx and the Post-Modern*. London: Zed Books.

Samuels, W. J. 1972: Welfare economics, power, and property. In G. Wunderlich and W. L. Gibson (eds.), *Perspectives of Property*. Institute for Research on Land and Water Resources, Pennsylvania State University, 61–148. Reprinted in W. J. Samuels and A. A. Schmid (eds.) 1981: *Law and Economics: An Institutional Perspective*. Boston: Martinus Nijhoff, 9–75.

Samuels, W. J. 1992: *Essays on the Economic Role of Government*. Vol. I: *Fundamentals*; Vol. II: *Applications*. London: Macmillan.

Samuels, W. J. and Schmid, A. A. (eds.) 1981: *Law and Economics: An Institutional Perspective*. Boston: Martinus Nijhoff.

Schmid, A. A. 1978: *Property, Power and Public Choice*. New York: Praeger.

Sen, A. 1970: The impossibility of a Paretian liberal. *Journal of Political Economy*, 78, 152–7.

Stone, C. D. 1987: *Earth and Other Ethics: The Case for Moral Pluralism*. New York: Harper & Row.

World Commission on Dams 2000: Report. Available at http://www.damsreport.org/

World Commission on Environment and Development 1987: *Our Common Future*. Oxford: Oxford University Press.

PART V

Ethics in Action: Empirical Analyses

CHAPTER 13

Empirical Signs of Ethical Concern in Economic Valuation of the Environment

Clive L. Spash

Some claim that consumer theory does "not exclude *a priori* any individual ethical system" and is "philosophically and psychologically neutral" (Malinvaud, 1972). In fact, this theory is based on a philosophy of preference utilitarianism and a restricted, largely hedonistic, model of psychological behavior. Nonconsequentialist reasoning is ignored, although it has proven relevant in environmental valuation (Spash, 1997). Environmental philosophers have emphasized the policy relevance of refusals to make tradeoffs on ethical grounds (O'Neill, 1993; Holland, 1995), but economists struggle to explain even simple refusals to consume market goods. For example, the rejection of tobacco by a nonsmoker would be assumed to disappear if the compensation offered were high enough. Continued refusal to accept such a payment would be regarded as merely strategic behavior. This has led to certain forms of common behavior being deemed incomprehensible or irrational within mainstream economics.

Applications of the contingent valuation method (CVM), to place monetary values upon aspects of the environment, have confronted economists with the inadequacy of their model of human behavior. From a psychological perspective, the CVM attempts to obtain a statement of intended (as opposed to actual) behavior, such as willingness to pay (WTP) or willingness to accept (WTA). An understanding of the behavior of respondents requires the inclusion of attitudes, ethical beliefs, and social norms as reasons for undertaking an action. Instead of investigating such motives for action, economic research on the CVM has emphasized the linking of stated preferences with actual behavior. Thus, economists struggle to explain survey results where, for example, people protest by bidding zero although they value the environment and can afford to pay for its protection.

This has created controversy over the interpretation of results, particularly when litigation is involved. Litigation in the US led the National Oceanic and Atmospheric Administration (NOAA) to commission an expert panel to produce a set of rules for conducting the CVM (Arrow et al., 1993). However, these rules have restricted original research on psychology and economics using the CVM. Variations in research design from "best practice" are susceptible to attack as "failing to conform to the rules." A prime example is the stance of the NOAA Panel on willingness to accept compensation. Variations in WTP and WTA for the same "good" can be large, because the underlying motives relate to both property rights and the psychological difference between paying and being paid compensation. Despite acknowledging that WTA is the theoretically correct measure for damage assessment, the NOAA Panel recommended the universal use of WTP as a "conservative" estimate. Strong criticism of this decision has failed to impact peer practice (Knetsch, 1994).

The desire to be "conservative" has also been imported to the UK. For example, in 1999 the largest CVM survey ever conducted in the UK (total sample size 10,650) was completed to provide evidence for a possible aggregates tax (Department of the Environment, Transport and the Regions, 1999). Monetary estimates of environmental damages were obtained on the advice of a CVM expert panel (Ian Bateman, Nick Hanley, Michael Hanemann, Susana Mourato, Richard Ready, and Ken Willis), using "an approach that is more likely to produce conservative results" (1999: 12). Amongst the "conservative" design principles was the incorrect welfare measure for local damages (WTP instead of WTA) and aggregation of the final results using a 25 percent discount rate (1999: 36).

While rules set by expert panels can become restrictive, and be used out of context, there are more general design features in any CVM survey that are desirable. These include clear description of the institutional context, explaining the consequences and expected benefits of choices, acknowledging how survey design can lead to or stimulate a given response, and generally producing a realistic scenario. Without addressing such issues, CVM research will fail to be relevant to the policy debate that surrounds it and may be judged inadequate or misleading with respect to the scientific analysis of public perceptions.

Within the context of these concerns, research has been produced, by or in cooperation with psychologists, which claims to show a motivational basis for WTP that diverges from that being assumed by economists. Thus, WTP has been described as the purchase of moral satisfaction rather than a trade or exchange value (Kahneman and Knetsch, 1992) and this has been linked to a contribution model of giving (Kahneman et al., 1993). In addition, psychological research into the motives behind WTP has incorporated environmental attitudes and/or norms (Stern, Dietz, and Kalof, 1993; Guagano, Dietz, and Stern, 1994; Stern et al., 1995). One conclusion has been that WTP may be only a measure of environmental attitudes (Kahneman et al., 1993; Guagano, Dietz, and Stern, 1994).

In what follows, the first section reviews the work of Guagnano, Dietz, and Stern (hereafter GDS) as an example of how misleading conclusions have been drawn. Despite the criticism, the social psychologists' work identifies the need for economic research on behavioral motives. The second section turns to the analysis of attitudes with particular regard to egoistic, social altruistic, and biospheric orientations. Some of my own work on extending and improving the analysis of refusals to trade in economic models is presented in the third and fourth sections, which explore the relationship between an individual's ethical stance and his or her stated intention to pay for an environmental change. The chapter concludes with some more general observations arising from the discussed research.

Social Psychology and CVM

In this section, the critique by social psychologists of economic valuation work is explored. Economists rely upon a model of behavior that assumes that values result from a given preexisting preference ordering, and are merely articulated during a survey to reveal a "true" value (see Kask, Shogren, and Morton, 1997). In contrast, psychology currently favors a theory of constructed preferences which are formed as required; for example, during the survey process. While preferences may be constructed, stable attitudes can still exist. As Kahneman et al. (1993: 310) state, the psychological approach "emphasizes the lability of preferences and their susceptibility to framing effects and to variations in context and elicitation procedures."

This difference in perspectives extends into the interpretation of WTP results. Kahneman et al. (1993) argue that economists interpret WTP responses as purchasing a public good, but that respondents are in fact making charitable contributions. The contrast is between buying to receive a range of benefits as opposed to merely supporting a good cause. The WTP measure may then become a surrogate for attitudes toward an environmental problem and more conventional psychological measures of attitudes could be substituted. In this case, WTP fails to represent either (i) the purchase of benefits or (ii) a stated intention to undertake a specific action (behavior). Under a reasoned action model, attitudes precede an intention to act, and combine with social norms and ethical beliefs to determine behavior (Fishbein and Ajzen, 1975). Thus, Kahneman et al. (1993: 314) regard economists' preoccupation with truth and strategic deception to substantiate validity in CVM as calling upon inappropriate categorizations of behavior.

The charitable contribution model may provide insights into why varying the magnitude of benefits often has little impact on the stated intention to pay. The respondents may focus on the basic problem that remains unaffected by variations in the size of the issue. In providing evidence of this phenomenon (termed "embedding"), Kahneman and Knetsch (1992) link the behavior to the purchase of moral satisfaction. Payments may be invariant with respect to specific consequences because the individual is only making a charitable contribution for the sake of their own benefit, a "feel good" factor. This type of behavior has been termed a "warm glow effect" (Andreoni, 1989). Other interpretations are also possible, including a nonconsequentialist ethical stance for which varying the outcome, such as the number of birds covered in oil, is less important or irrelevant in comparison to the type of action (causing harm to animals). However, research into the underlying ethical motives has been neglected, while the concept of embedding has been open to speculative attack. For example, the NOAA panel stated that the idea of WTP falling to zero for additional environmental

Table 13.1 An analysis of framing of the WTP question in Guagnano, Dietz, and Stern (1994)

Assumed frame	Environmental "good"	Bid vehicle	WTP question
"Consumer good"	Greenhouse effect	Gasoline (implicit tax?)	"Burning fossil fuels is believed to be one of the main contributing factors to global warming, sometimes called the greenhouse effect. It's been suggested that raising gasoline prices would substantially reduce the use of fossil fuels. Assuming that would work, how much extra would you be willing to pay for a gallon of gasoline to help reduce global warming?"
"Consumer good"	Paper recycling	Paper towels (?)	"At most grocery stores, paper towels cost about 85 cents per roll. How much extra would you be willing to spend for a roll of paper towels made from recycled paper products?"
"Contribution"	Biodiversity, deforestation	International trust fund	"Scientists are becoming increasingly concerned about the loss of many species of animals in Latin America due to heavy tree cutting in the rain forest. If the wealthier nations of the world, including the United States, were asked to establish a fund to preserve these forests, how much would you be willing to contribute to a one-time fund of this type?"
"Noncontribution"	Biodiversity, deforestation	National (?) tax	As above, but with the wording: "What do you think would be a reasonable dollar amount for your taxes to increase to solve the problem?"
"Contribution"	Human health, water quality	Local (?) trust fund	"Some people are concerned that increasing amounts of toxic chemicals are making their way into our drinking water. In the event that one of these chemicals was found in the Fairfax County water supply and no responsible party could be identified, what would you be willing to contribute to a one-time fund to solve the problem?"
"Noncontribution"	Human health, water quality	National (?)/local (?) tax	As above, but with the wording: "What do you think would be a reasonable dollar amount for your taxes to increase to solve the problem?"

protection was ". . . hard to explain as the expression of a consistent, rational set of choices." These experts regard such individuals as irrational, which has led to the rejection of responses by certain kinds of individuals from data analysis and policy advice.

If the aim is to obtain a rank-order of issues, Kahneman et al. (1993: 314) conclude that "WTP is not the preferred way of doing so because it is psychometrically inferior to alternative measures of the same attitude." Yet they hold back from extending their results to the CVM, because their survey design was unconventional (it lacked information content and presented multiple issues for valuation). In addition, GDS (1994: 411) felt that this work "did not specify nor directly test a contribution model." They aimed to rectify this situation and engage with the economists' interpretation of CVM results. However, the approach taken by GDS is overly complex and confuses several issues.

GDS sampled 367 members of the general public in Virginia by phone. A variety of WTP questions were employed in order to separate out contributory giving, as summarized in table 13.1. Four environmental "goods" were included in the study: reduced global warming, increased paper recycling, reduced deforestation, and cleaning-up chemical contamination of local drinking water. For two of these "goods" there were two alternative payment mechanisms (trust or tax), giving a total of six WTP scenarios.

Problems arise in the classification of these six scenarios, on the basis of the payment method, as either consumer (gasoline, paper towels), contributory (trust fund), or noncontributory (taxes). First, all scenarios have public goods characteristics, which makes the consumer good category a misclassification. All willingness-to-pay responses relate to changes in environmental quality, which are public goods. Second, while the tax scenarios are supposed to represent a noncontributory frame, this logic is not extended to the scenario for preventing global warming, where a tax could be inferred.

In general, it seems uncertain how the contributory model should be represented. GDS initially equate it with being willing to pay for a public good, and a "purchase model" is equated with buying a private good. In contrast, Kahneman et al. (1993) regard the contribution model as operating where giving is to support a "good cause." That is, they appeal to the respondents' interpretation of why they are giving rather than whether the good could be defined *a priori* as public or private. Indeed, public goods such as nuclear weapons may fail to be regarded as "good causes," and individuals may regard benefits of public goods (such as those communally provided) as being bought, as under the purchase model. The inadequacy of the public goods definition is apparent because of the need to impose a "contribution frame" (Guagnano, Dietz, and Stern, 1994: 412). This framing implies that different categories of giving are stimulated by changing the descriptive circumstances. More specifically, switching the method of payment from a trust to taxes is meant to correspond to a move in frame from contributory to noncontributory.

If the argument that the payment method determines whether a contributory model is operative is accepted, then taxes must be misclassified. Taxes are often a universal method of payment, spread throughout the community (all employees are assessed for income tax). This links payment to contributions by others and the community as a whole. In fact, income taxes have been recommended in the US as neutral bid vehicles for use in CVM surveys. In contrast, a payment to trust allows people to free-ride. There appears to be no clear reason for a trust to represent a contributory frame and a tax a noncontributory frame.

If the analyst could switch on or switch off a contributory versus a purchase model by selecting the payment mechanism, the solution for CVM practitioners would be simple. The contribution model has relevance because it relates to respondents' regard for the type of object – not the method of payment. As stated by Kahneman et al. (1993), "The impetus for charitable giving is the urgency of the problem, not the attractiveness of the solution. Accordingly, we expect participants in WTP surveys to focus on problems, and to show little sensitivity to interventions."

Besides the above points, there are two broad reasons for being critical of the extent to which the study is representative of economic work on the CVM. First, the hypothetical market is unrealistic, because respondents are ill-informed as to what they are purchasing, how it will be provided, and how they will pay (as table 13.1 shows). Second, the descriptions introduce unmeasured variability into the scenarios. This extends to the implied institutional contexts, and the responsibility for the environmental problems to which the scenarios allude. Thus, the work appears to suffer the same lack of comparability to CVM as Kahneman et al. recognized in their own work, but neither is without insight.

Motives for Giving: Attitudes and Altruism

The most interesting aspect of social psychological work relating to CVM is the use made of attitudinal scales related to altruistic and biospheric values. GDS hypothesized that altruism should be related to contributory giving, and that "Willingness to pay higher prices for environmental goods is viewed as altruistic behavior because the extra money people pay provides environmental benefits that are public goods" (Guagnano, Dietz, and Stern, 1994: 412). That the type of good (public versus private) has a relationship to altruism cannot be tested here, because there were only public goods in this study. Thus, it should not be a surprise that in the case of the "consumer goods" the authors found that ". . . decisions about these goods incorporate an element of altruistic concern" (ibid.: 414).

GDS employed attitudinal scales that addressed moral norms based upon asking for agreement or disagreement with statements, or items, which were then aggregated into scales for use in statistical analysis of WTP (but only positive bids). The study used a two-item scale on "ascribing responsibility" (AR) to oneself for ameliorating environmental problems and a three-item scale on "awareness of negative consequences" (AC) for others (human and nonhuman). These items can be related to biospheric and social altruistic values, as discussed in the next section. A third two-item scale measured perceived personal costs (PC) as an indicator of "self-interest calculations" (ibid.: 412).

The results showed that AC was significantly related to WTP under three scenarios and AR under four, while the PC scale was insignificant across all cases. Since all of the environmental problems were public goods, the positive influence of both the AC and the AR scales on WTP should have occurred across all six scenarios. The AC and AR scales were both related to WTP for the trust scenarios to clean up chemicals and prevent deforestation. The AC scale was also positively and significantly found to determine WTP taxes to prevent deforestation, while the AR scale explained WTP for reducing global warming and increasing recycling. Both AC and AR were insignificant predictors of WTP taxes for cleaning up chemicals, but this model also failed the F-test. This was the only model where income was significant, which the authors took to mean that other factors were more relevant than ability to pay in all WTP cases, although this would also result from the general failure to specify what was being purchased.

Altruism was predicted to be an important determinant of WTP and so to provide support for the contributory interpretation of giving. However, the study was unclear as to whether altruism was expected to determine WTP for all public goods or only those framed as contributory. The authors concluded: "Our findings show that stated willingness to pay extra taxes to achieve environmental protection does not follow a model of altruistic behaviour." The fact that AC showed a positive significant influence on the (supposed noncontributory) scenario of taxes for forests contradicts both this claim and also that altruism was driven by the "contributory frame." In addition, as explained above, the theoretical case for altruism being excluded by the use of taxes as a bid vehicle seems weak. On the basis of this study, Stern et al. (1995: 1631) have stated that:

> When contingent valuation items were framed as contributions to a fund to support environmental protection, willingness to pay was strongly influenced by beliefs about consequences of environmental degradation, but the effects disappeared when the questions were framed as willingness to pay taxes for the same environmental protections.

Yet, the scenarios were not strictly identical due to differences in and lack of specificity concerning payment, institutional context, and level of decision-making (local, national, or international). The conclusion also ignores the fact that the AC variable remained highly significant for the tax to preserve forests in Latin America. If the most significant coefficients are considered ($p < 0.01$), then the relationship between AC and WTP is seen to be strongest for deforestation, regardless of the payment mechanism. This would seem to support the contention of Kahneman et al. that the environmental problem provides the focus under charitable giving. However, support for the contribution model then comes from insensitivity to framing, rather than being caused by it.

The results of GDS (1994) also require some reconciliation with related research. The AC scale contains items that have appeared in other attitudinal scales: two representing biospheric values and one representing social altruism (Stern, Dietz, and Kalof, 1993; Stern, Dietz, and Guagnano, 1995; Stern et al.,

1995). The two items in the AR scale can also be viewed as representing biospheric and social-altruistic values respectively. The two-item PC scale was previously used by Stern, Dietz, and Kalof (1993) as a part of a three-item measure of egoistic beliefs about consequences and found to be a significant determinant of WTP. That study used two very general payment scenarios: a request for payments by income tax and a gasoline price rise to "protect the environment." Stern, Dietz, and Kalof (1993: 336) stated:

> Questions about willingness to pay draw respondents' attention to the things on which they spend money, and these things are more likely to pertain to their well-being than to social-altruistic or biospheric value. If this argument is correct, a willingness-to-pay question has the effect of focusing attention on the egoistic value orientation.

However, in GDS (1994) the egoistic PC scale was insignificant across all six WTP questions, while the biospheric–altruistic AC scale was significant for three WTP cases. In addition, the AR scale, which concerned the protection of other species (biospheric) and other people (social-altruistic) was significant in four WTP scenarios. Thus, five out of the six WTP scenarios showed a significant influence with regard to biospheric and social-altruistic motives, and none with regard to egoistic motives.

The unspecified nature of the WTP questions in GDS (1994) may have encouraged shows of altruism, because there was no obvious or explicit direct personal gain to the respondent from payment. This would explain the failure of the egoistic measure – the only surprise then is that a stronger relationship with AC across all six WTP questions was absent. In Stern, Dietz, and Kalof (1993), the use of a student sample may help to explain the difference in results. However, while there is a relationship between the students' egoistic attitude and WTP, the biospheric AC scale was also significant in one of the two WTP cases (payment by income tax). Thus, there does seem to be consistency concerning the role of nonegoistic motives.

In summary, CVM surveys may reveal a range of motives besides the purely egoistic consequential perspective that is normally assumed to be dominant in economic studies. The charitable contribution model and the embedding effect literature have alluded to moral satisfaction as a motive, but without specific investigation or elaboration. In addition, the range of work by social psychologists such as Dietz, Guagnano, Kalof, and Stern, as well as that by Kahneman and his co-authors, has failed to incorporate hypothetical environmental tradeoffs with sufficient detail to approach the practice of CVM studies.

Biodiversity in Jamaica and Curaçao

This and the next section report on two empirical analyses of ethical motives for WTP under a standard economic approach to CVM. The first was conducted for The World Bank and addressed coral reef biodiversity degradation (Spash, 2000).

The second related to a wetland re-creation scheme in the UK (Spash et al., 1998). The surveys were designed in several sections, which were delivered to respondents in the following order: framing and knowledge questions, the information pack and payment scenario (open-ended WTP), ethical and attitudinal questions, and socioeconomic data.

The World Bank study investigated whether the CVM could be used to assess the benefits of maintaining and improving coral reef biodiversity for Curaçao and Jamaica (Spash et al., 1998). The sample sizes of these surveys were 1,152 and 1,058 respondents, respectively. Among the methodological issues of concern was the refusal to trade by those giving zero bids as described by lexicographic preferences. Such preferences arise when goods are ranked in an absolute ordering as in a lexicon. Standard economic models regard this as an uncommon subcategory of preferences, although in marketing the concept is more commonly accepted. For example, a person may always prefer a blue car to a pink car, regardless of any other features the car may have as extras. The literature demonstrates that such preferences can be common and create problems for the interpretation of CVM results (Spash, 1997, 1998).

In addition, motives underlying positive bids proved to be of interest, because they too can be consistent with lexicographic preferences, or given for reasons that conflict with economic assumptions. The surveys took a rights-based ethical position as signifying an ethical stance compatible with lexicographic preferences. That is, some people may hold strong beliefs in rights that prevent their making any tradeoffs (for example, a belief in animal rights), and they then rank options by their ethical attributes.

The Montego Bay Marine Park in Jamaica provided an actual institution with a record of marine ecosystem management and a realistic context within which a WTP scenario could be developed. In Curaçao, a marine park along the whole southern coast was being planned and was used in the CVM. Environmental quality within the proposed parks was characterized to give a background picture and projected trends in coral biodiversity. A status quo scenario for the parks resulted from a literature review and expert advice. This helped to summarize the current situation in terms of coral reef quality and causes of degradation. Two states of the coral reef were then relevant: the current degraded condition and a healthy coral reef under management options.

Due to variations in coral reef degradation, the level of improvement in coral abundance expected from the management options differed between the two countries. The park proposed for Curaçao was much larger than that for Jamaica, while the increase in biodiversity was lower (changing from 50 percent to 75 percent, as opposed to 75 percent to 100 percent). Information on physical changes was summarized using color maps, descriptions read aloud by the interviewer, and show-cards. The identification of causes of reef degradation was combined with knowledge of the powers and jurisdiction of park institutions to simultaneously determine the type of management options that could realistically be included in the survey.

The samples for each study included both tourists and locals. Respondents were asked to contribute toward a trust fund managed by the marine park in order to increase marine biodiversity within the park boundaries. The payment was to be on a per annum basis for five years. The technique for elicitation of WTP was an open-ended question that was chosen as being straightforward and realistic. The environmental improvement being purchased was a rise in marine biodiversity within the areas by 25 percent, which was contrasted with a status quo stability scenario and a no-management scenario that would cause a 15 percent reduction in biodiversity.

The results showed that 50 percent (574) of the Curaçao and 64 percent (680) of the Jamaican respondents had a positive WTP. Three reasons normally regarded as consistent with economic theory (lack of income, improvement unimportant, and other goods more important), accounted for 46 percent and 41 percent of zero bids respectively. A lack of income proved to be the largest category and was disproportionate in relation to the socioeconomic profile of the samples. In addition, some tourists felt that this was "not my problem," but that they would contribute to a similar scheme in their own country (39 percent of tourists in Curaçao and 21 percent in Jamaica). This was classified as a zero bid for the reason of zero value, although – when probed – some respondents did state that they would be willing to pay a user fee for direct benefits. The remaining zero bids were various protests, including: "free-riders" (only 1–2 percent), those who felt that paying was an inadequate solution, some who had a lack of faith in the proposed marine park and trust fund, and those who rejected the payment mechanism. In Curaçao, the latter expressed a general feeling that the marine park trust should be a government responsibility, and that taxes were already very high. Thus, even if the design had used a tax payment mechanism, the protest bid would have persisted and may have been larger. Overall, many of the respondents valued biodiversity but refused to give a positive WTP amount. This is of concern, given that 32 percent and 27 percent of zero bids for Curaçao and Jamaica, respectively, fell into these four categories. As seen above, most work (for example, GDS) has concentrated solely on positive bids.

Respondents were asked to state the extent to which they saw rights to protection from harm as operating (absolute, circumstantial, or irrelevant) in relation to each of five categories (present humans, future humans, marine animals, plants, and ecosystems). Almost all of the sample were prepared to attribute absolute rights to current and future humans. Marine animals, plants, and ecosystems were also attributed absolute rights by approximately 60 percent of the Curaçao sample and over 80 percent of the Jamaican sample. Respondents could answer that they just "did not know," but only 0.2 percent in Jamaica and 2.1 percent in Curaçao found this necessary.

The respondents who had attributed any rights to one of the five categories were next asked whether, in the case of the relevant marine park, they believed that the rights they had attributed meant a personal responsibility to prevent harm, regardless of the cost. Approximately 79 percent of the Jamaican and 68 percent of Curaçao sample answered in the affirmative. Those who affirmed that they had a personal responsibility, regardless of the cost, were asked whether they would accept harm to the relevant island's marine life and habitat if attempts to prevent it would threaten their current standard of living. The other group of respondents – who had denied rights in this case – were also asked to reconsider, given a more specific scenario including a threshold personal impact. In their case, they were asked whether they would accept a personal duty to avoid harming the relevant island's marine life and habitat if their current standard of living would remain unaffected. The outcome of these questions was to enable the sample to be split into four categories, as follows:

1   Those who attributed rights and accepted a strong personal responsibility to protect marine life and habitats from harm even when their standard of living was threatened.
2   Those who attributed rights and accepted a personal responsibility to protect marine life and habitats from harm only if their own current standard of living was unaffected.
3   Those who withdrew rights and any personal responsibility to avoid harm to marine life and habitats when the cost of doing so was in terms of their current standard of living.
4   Those who rejected rights and any personal responsibility to protect marine life and habitats from harm, regardless of whether or not their own current standard of living was affected.

In addition, there were those who rejected rights in general, rather than in this particular case, who formed a minority fifth category.

The results showed a dramatic reduction in those attributing absolute or strong rights (category 1 above), from 79 percent down to 14 percent for Jamaica and from 68 percent down to 27 percent for Curaçao. The two middle categories, 2 and 3 above, show a threshold effect that might be consistent with a modified lexicographic position. That is, once a basic standard of living is obtained, a stronger ethical position is adopted with regard to other species (Spash, 1998). A readiness to consider the tradeoff circumstances and the subjectivity of the relevant standard of living mean that individuals in these categories may be regarded as acting as consequentialist and weighing up the tradeoffs. This study left the distinction between the consequential and these weak rights positions indistinct, and this was rectified when the wetlands study was conducted, as discussed in the next section.

One hypothesis was that individuals' actions in protesting against CVM and bidding nothing could be explained in part as holding and defending rights and/or duties. The survey allowed for bids by both time and money, which reduced the zero-bid category beyond monetary WTP. The zero bidders as a subgroup of strong duty-holders then only accounted for 3.4–7.5 percent. Of these, respondents who gave a protest reasons for refusing to pay accounted for 2.9 percent of the Curaçao sample and 1.7 percent of the Jamaican sample. The result was similar across tourists and locals. Thus, strong duties explained 15 percent and 11 percent of all the protest bids; that is, refusals to pay for reasons of nonzero value.

All strong duty-holders were asked how they expected environmental rights to be protected within the marine parks. In Jamaica 66 percent (10 percent of the total sample) and in Curaçao 48 percent (13 percent of the total sample) wanted either a legal approach or education, or a combination of the two. Some of those holding a strong duty position felt that the trust fund was also a good idea, and would help in the protection of the rights that they had attributed to the marine environment. Those who held a strong duty position and protested in terms of a zero bid also favored legal and educational approaches. In Jamaica, 50 percent of these individuals opted for a purely legal approach; while in Curaçao, 53 percent wanted either a legal and/or an educational approach. Thus, focusing on the issue was compatible with desiring specific institutional arrangements, which is somewhat contrary to the contribution model.

A set of variables measuring different aspects of the ethical stance being taken by the respondent was included in a bid-curve analysis, using a semilog-linear form. For Curaçao, the determinants of WTP were a standard set of socioeconomic variables (sex, age, and education), knowledge, and the positions taken toward rights. Income was correlated with age and education, and suffered item nonresponse (only 642 responses). Knowledge of marine biodiversity and the direct use variables proved to be positive and significant determinants of WTP. A seven-point scale was designed to capture attitudes toward the attribution of a right to be protected from harm to marine animals, plants, and ecosystems. Those attributing absolute rights to all three aspects of the marine environment were ranked highest, and those denying rights in all three cases were ranked lowest, with a graduating scale between these two extremes. Rights for the marine environment were positively related to WTP. The role of ethical positions was further confirmed by the significance of dummy variables on the personal duty to protect the life and habitats of the marine park; that is, respondents taking a strong duty perspective or rejecting any duty. A strong personal duty regardless of the cost was positively correlated with WTP, while the rejection of this duty reduced WTP. A variable on the difficulty found with the section on ethical questions proved significant and positively correlated with WTP. This may mean that those concerned about biodiversity improvement struggled with their precise ethical positions and the extent to which duties were for them weak (tradable) or strong (lexical). Overall, the results for Curaçao show a model of WTP being dependent upon standard socioeconomic variables plus rights and duty-based variables.

A similar model was run for Jamaica, including a set of variables covering socioeconomic status, knowledge, and the position taken toward rights. A dummy variable for tourists versus locals was strongly significant and negatively correlated with tourists. The knowledge and use variables again proved to be positive and significant determinants of WTP. In Jamaica, the set of variables on ethical stance were less relevant. However, the role of ethical positions was confirmed by the significance of the dummy variable rejecting any duty. This was also negatively correlated with WTP, as was the case for Curaçao. The overall results for Jamaica were in line with those for Curaçao, except in that the model lacked the range of significant rights and strong duty variables. While the model was weaker in terms of predictive power, with the exception of gender all of the variables in the model were significant at the 99 percent level.

Species Rights in the UK

In this study, a small site (one square mile) in eastern England, currently used for crop farming, was hypothesized as being purchased by an existing regional charity concerned with the conservation of wetlands. A request was made for a one-off payment to a trust fund established specifically for the project. An information pack was designed, consisting of an area map, photographs of an actual site before and after conversion to a wetland, an artist's impressions of the two ecosystems, and brief descriptions. The wetlands and agricultural scenarios were referred to as different potential uses of the area without any specification as to which might be preferable.

The sample size was 713 and approximately a third of respondents gave a positive WTP. Three categories of people, totalling 466 respondents, gave no monetary valuation but might hold a positive value for the environmental change. These were zero bidders, refusals, and don't knows. There were 36 respondents who refused to answer the WTP question and 182 who responded "don't know." Standard reasons regarded as legitimate explanations for bidding zero (low income or finding the change unimportant) accounted for 286 respondents.

In order to categorize ethical positions, respondents were told that: "A major aim of re-creating wetland is to provide sanctuary for endangered species of birds such as Bewick's swan, the pintail and gadwall." Respondents were then asked to match one of four motives (rights for animals, consequentialism in a preference utilitarian mode favoring either endangered species or humans, and superiority of humans) with their reasoning for their response to the WTP question. Those attributing rights to bird species were then confronted with a scenario of a personal cost that reduced their standard of living to what they regarded as a minimum. Under such circumstances, the respondents were asked whether they would still protect the birds' right to life, or accept that some bird species might become extinct. A category of strong rights is consistent with lexicographic preferences, and this connection is discussed in a more comprehensive report of the results (Spash, 2000). Those who backed down when confronted with the personal cost scenario were taken to hold a weak expression of rights. In both national and 'ocal samples, a larger number maintained their position (strong rights) than accepted species extinction, and the proportions in each category were similar. This process gave five ethical categories: strong rights-based, favoring endangered species even when personal living standard was reduced to a minimum; weak rights-based, relinquishing rights if threatened with a personal cost that would reduce living standard to a minimum; consequentialist favoring species; consequentialist favoring humans; and a human priority position, where humans come first regardless of the consequences.

Results for the entire sample, including "don't know" answers to the ethical questions, showed that 37 percent attributed rights to birds, 9 percent put humans first, and only 47 percent weighed up the consequences of the case (in accordance with economic theory). Among the 180 protest nonbidders, 76 held the two rights-based positions, giving 11 percent of the total sample as showing behavior consistent with a rights motive and protesting against bidding. Those who were regarded as "legitimate" nonbidders for the purposes of a standard CVM study, because they failed to give a suitable protest reason, should be recognized as potentially holding a position that is inconsistent with economic assumptions. This applies particularly to those who claimed an income constraint. In particular, 4 percent of the total sample did so under weak rights and 5 percent under strong rights.

In addition to nonbidding strong rights-holders, 15 percent of the total population sample held either strong or weak rights while they decided to bid positively. Such positive bidding could represent consistent behavior for those with a weak right, where they contribute a fixed amount that they regard as meeting a threshold. Alternatively, the behavior may be regarded as inconsistent with the statement that endangered bird species have the right to protection, because a monetary value is now being placed upon the project in order to achieve that protection. Either way, the motivation behind the WTP seems to conflict with regarding the monetary value as an exchange price or a compensatory payment.

Bid-curve analysis using a semilog-linear function showed the significance of all of the ethical positions, including the consequentialist. There were 495 positive and zero bidders in the sample, which was reduced to 458 by item nonresponse. Education and gender were used as surrogates for income data due to refusals to answer and under-reporting. Variables that covered the likelihood of visiting the wetland site in the future, environmental awareness, and education to 16 years of age all proved highly significant. The model was significant on the F-test and had an adjusted R^2 of 23.5 percent – which is high for CVM studies, where a value of 15 percent is an acceptable level. All of the variables were significant at the 90 percent level, and the ethical variables at the 95 percent level. One of the most highly significant variables was the strong rights position. Both the variable for consequentialists favoring birds in the case of the wetlands project and that for those who placed humans first regardless of the circumstances were significant at the 98 percent level.

Conclusions

The research on coral reef biodiversity showed that the WTP of both locals and tourists was related to their ethical position on rights. A positive bid for an environmental improvement proved to be positively related to the belief in duties toward environmental entities. The monetary amounts stated included expressions of multiple values, some of which related to the moral concern to protect marine animals, plants, and ecosystems. The pricing of all aspects of the marine environment as just another commodity will then fail to reflect the rich range of values that individuals associate with their environment, and the meanings that they associate with their bids.

The results concerning the income-constrained rights-based categories are important because of the way in which CVM practitioners tend to differentiate their treatment of nonbidders by protest category. Protest nonbidders may be treated identically to zero bids or they may be given an imputed bid (the mean WTP of positive bidders). Thus, drawing the boundary line between these categories can be crucial to the resulting aggregated WTP estimate. If applied to the rights respondents, either of these standard treatments of protest nonbidders would seem inconsistent with the values being expressed.

Overall, the results from the standard CVM survey approach to WTP question design support the concerns in the social psychology literature about the importance of a range of motives for giving. The studies reported here show the relevance of different ethical positions besides the consequentialist, which environmental economists assume to be universal. One result is that WTP reflects nonexchange values and cannot therefore be regarded as commensurate with market prices. Human value formation with respect to the environment appears to be far more complex than economists have previously assumed, and combines both attitudes and ethical and economic values. Interaction effects between causes, motives, and behavior will then help to explain the variety and meaning of responses. This need for explanation is necessary to counter claims that respondents who act outside the economic model are "irrational" and that their stated (or actual) behavior is inexplicable. Rather than seeing the challenge as how to downplay, separate, and remove their values from the policy process, the aim should be to consider how values that are apparently "noneconomic" can be included.

The hypothesis that a contribution model means insensitivity on behalf of respondents to certain aspects of framing, such as payment mechanism, has mixed empirical support. While results from GDS show insensitivity to payment mechanism, those from the coral reef survey show concern by respondents for the institution and mechanism of payment. In the latter case, the respondents were apparently focusing upon the environmental attribute, possibly as a charitable donation, but also on the "attractiveness of the solution."

The psychological research into environmental charitable giving is interesting, but the results are mitigated by the poor specification of the scenarios. This may encourage a biospheric-oriented individual to bid positively if he or she can see some prospect of positive consequences for the environment (and believes in the institutional context). Simultaneously, an individual's egoistic attitudes may produce a positive intention to pay, if he or she can gain moral satisfaction from giving to a good cause. Either way, the psychological research bears only a tangential relationship to CVM studies, which are grounded in the welfare theory of neoclassical economics. A CVM study requires a well-specified environmental change and institutional context. Overall, the evidence provided needs clarification and reinterpretation, and then does seem to support the role of nonegoistic attitudes as motives for WTP.

This chapter also shows that there is evidence of nonconsequentialist, and potentially biocentric, reasoning in answers to WTP under the standard approach to CVM surveys. In the wetlands survey, there was a positive correlation between the rights positions and WTP, and a negative one for those favoring humans above all else. Half of those who gave a positive bid attributed rights to endangered bird species and so readily identified their motives with nonconsequentialist reasoning. This extends the concern over the values being derived by the use of CVM surveys from the misclassification of protest bidders, who may hold noncompensatory preferences, to the motives behind and meanings of the positive bids.

One implication of this work is the need to develop alternatives to monetary valuation as an environmental policy tool. The oft-cited case is that there are "no options," so that despite all the faults in benefit–cost and economic analyses, these tools must be employed. This is blatantly false. There is a range of methods under multiple criteria analysis, some of which may subsume monetary values but can allow for incommensurability. Political scientists have been developing participatory approaches for deliberating on environmental issues (citizens juries, consensus conferences, and so on). In development studies, techniques such as rapid rural appraisal have been applied, along with various deliberative approaches. Local environmental activists and planners have employed "planning for real." Thus, researchers and government agencies have a considerable range of possible tools for different types of issues, but these require breaking away from the simplistic reductionism of the audit culture, found in modern management, and acceptance of the complexity that is inherent in environmental and social systems.

References

Andreoni, J. 1989: Giving with impure altruism: application to charity and Ricardian equivalence. *Journal of Political Economy*, 97(6), 1447–58.

Arrow, K., Solow, R., Portney, P. R., Leamer, E., Radner, R., and Schuman, H. 1993: *Report of the NOAA Panel on Contingent Valuation*. Washington, DC: Resources for the Future, 38.

Department of the Environment, Transport and the Regions 1999: *The Environmental Costs and Benefits of the Supply of Aggregates: Phase 2*. London, Department of the Environment Transport and the Regions, 208.

Fishbein, M. and Ajzen, I. 1975: *Belief, Attitude, Intention and Behavior: An Introduction to Theory and Research*. Reading, MA: Addison-Wesley.

Guagnano, G. A., Dietz, T., and Stern, P. C. 1994: Willingness to pay for public goods: a test of the contribution model. *Psychological Science*, 5(6), 411–15.

Holland, A. 1995: The assumptions of cost–benefit analysis: a philosopher's view. In K. G. Willis and J. T. Corkindale (eds.), *Environmental Valuation: New Perspectives*. Wallingford, UK: CAB International, 21–38.

Kahneman, D. and Knetsch, J. L. 1992: Valuing public goods: the purchase of moral satisfaction. *Journal of Environmental Economics and Management*, 22(1), 57–70.

Kahneman, D., Ritov, I., Jacowitz, K. E., and Grant, P. 1993: Stated willingness to pay for public goods: a psychological perspective. *Psychological Science*, 4(5), 310–15.

Kask, S., Shogren, J., and Morton, P. 1997: Valuing ecosystem change: theory and measurement. In J. van den Bergh and J. van der Straaten (eds.), *Economy and Ecosystems in Change: Analytical and Historical Approaches*. Cheltenham, UK: Edward Elgar, 291–312.

Knetsch, J. L. 1994: Environmental valuation: some problems of wrong questions and misleading answers. *Environmental Values*, 3(4), 351–68.

Malinvaud, E. 1972: Utility function and preference relation. In *Lectures on Microeconomic Theory*. Amsterdam: North-Holland, 19–20.

O'Neill, J. 1993: *Ecology, Policy and Politics: Human Well-Being and the Natural World*. London: Routledge.

Spash, C. L. 1997: Ethics and environmental attitudes with implications for economic valuation. *Journal of Environmental Management*, 50(4), 403–16.

Spash, C. L. 1998: Investigating individual motives for environmental action: lexicographic preferences, beliefs and attitudes. In J. Lemons, L. Westra, and R. Goodland (eds.), *Ecological Sustainability and Integrity: Concepts and Approaches 13*. Dordrecht: Kluwer, 46–62.

Spash, C. L. 2000: Ecosystems, contingent valuation and ethics: the case of wetlands re-creation. *Ecological Economics*, 34(2), 195–215.

Spash, C. L., van der Werff ten Bosch, J., Westmacott S., and Ruitenbeek, J. 1998: *Lexicographic Preferences and the Contingent Valuation of Coral Reef Biodiversity in Curaçao and Jamaica: Report to the World Bank*. Delft, Resource Analysis 135.

Stern, P. C., Dietz, T., and Guagnano, G. A. 1995: The new ecological paradigm in social-psychological context. *Environment and Behavior*, 27(6), 723–43.

Stern, P. C., Dietz, T., and Kalof, L. 1993: Value orientation, gender and environmental concern. *Environment and Behavior*, 25(3), 322–48.

Stern, P. C., Dietz, T., Kalof, L., and Guagnano, G. A. 1995: Values, beliefs and pro-environmental action: attitude formation toward emergent attitude objects. *Journal of Applied Social Psychology*, 25(18), 1611–36.

14

Motivating Existence Values: The Many and Varied Sources of the Stated WTP for Endangered Species

Andreas Kontoleon and Timothy Swanson[1]

What is it that individuals are valuing when they state a positive willingness to pay for the nonuse of some stock of an environmental amenity, such as an individual species? What are the various reasons that would cause them to be willing to pay to maintain stocks of animals that they do not use or experience individually? How can stock-related values exist within the economic model, which focuses on the maximization of individual flows of utility? Although economic theory tends to avoid the examination of the motives behind these economic values, it is increasingly being acknowledged that the concept of existence value should be examined (and perhaps defended) through the investigation of plausible motives for this residual component of total economic value (Loomis, 1988; McConnell, 1997).

   This chapter examines the various motivations behind the expression of positive existence values for endangered species. We term the primary motivations examined here *pure conservation* and *pure welfare*. The pure conservation motive is represented by the willingness to pay for the maintenance of the endangered species as a genetic stock, something that might be useful for future options or for the use of future generations. It is a motive based on self-interest, although the interest might be very remote in time and/or space. By contrast, the pure welfare motive is the willingness to pay for the enhancement of the welfare of the individuals of the endangered species itself. One motive is more closely

[1]   We would like to acknowledge the support of the China Council for International Cooperation on Environment and Development in the undertaking of the giant panda study, and the support of the Centre for Social and Economic Research on the Global Environment in the case of the black rhino study. We are grateful to the editors for extensive comments on an earlier draft. All remaining errors are our own.

related to the prospects for the survival of the species (a *quantity*-related motive), while the other is more closely related to the lifestyle experienced by individuals of the species while it continues to survive (a more *quality*-related motive). One of the objectives of the studies reported in this chapter has been to attempt to segregate between the willingness to pay for quantity versus quality in stated existence values. We will explain this further in the discussion below.

It is important to break down stated existence values into their component parts, because the nature of these parts, and their relationship to one another, is important. The motives set forth above are differentiable in important senses. They could engender similar responses to the same survey question (regarding the valuation of a stock of environmental amenity or species), while each re-spondent has in mind an entirely distinct – or even a competing – flow of value. For this reason, policies based on expressed existence values may in fact be conflating a wide variety of expressed preferences and viewpoints. For example, it is often the case that policies on endangered species are made by aggregating expressed preferences across wildly diverse human constituencies, ranging from vegans to hunters. It is important for policy purposes that we know the extent to which WTP can be aggregated across such divergent groups in society.

Equally important, it is necessary to examine existence value closely to enquire about its fundamental nature and meaning. Why do individuals report a positive willingness to pay for distant stocks of resources, rather than the flows of utility that they might receive from them? Is this indicative of a fundamental flaw in the economic model of individual welfare, or is it more indicative of a flaw in the conception of existence value?

In this chapter, we report briefly on a series of studies undertaken by the authors in which these issues have been investigated. We find that so-called existence values are probably best thought of as attempts by the respondents to channel flows of value to others about whom they care, rather than as a general willingness to provide stocks of the resource in the abstract. We believe that individuals are willing to pay to support policies that they believe will channel flows of value to other individuals and groups about which they care – even relatively remote groups, such as descendants or members of the endangered species itself.

If this is the case, then expressed stock-related values depend crucially on the expectations of the respondent about who will receive the benefits of the flows from those stocks. Individuals are well aware of the opportunity costs associated with channeling resources in one direction, as opposed to another. They view an expression of support for an endangered species as an expression of their support for the channeling of those resources in *their* desired direction. According to this view, their expressions of willingness to pay are equally expressions of withdrawal of support for the channeling of that value in other directions. Therefore, in order for a study on existence value to discern these tradeoffs, it should be conducted within the context of a policy choice experiment. Other-wise, individuals are not being allowed to indicate that which they wish to de-emphasize in order to provide their desired emphasis.

If respondents are asked only abstract questions concerning willingness to pay for stocks of natural resources, then they will answer these questions on the basis of their own preferred set of assumptions. This can lead to the unpalatable result in contingent valuation studies that individual statements of willingness to pay will be aggregated, when in fact they represent diametrically opposing preferences.

Allow us to motivate the entire exercise with an example. Every two years or so, the Convention on International Trade in Endangered Species (CITES) has a meeting of the parties to assess the manner in which the international community should take action in order to conserve endangered species. Although the meeting pertains only to the topic of policies for conserving species that are potentially endangered by trade, it often ends in rancor and controversy. For example, the parties first debated the management of the African elephant nearly 20 years ago, but they continue to do battle over the correct approach to the species even to this day (Barbier et al., 1990). This is because the range of motivations regarding endangered species covers everything from those interested in current and individual consumption of the goods and services that they can produce (the Japanese ivory carving industry, for example) to those interested in the current and future welfare of individual members of the species (for example, the memberships of the human societies of the world). If a survey on the existence value of the African elephant were undertaken on the floor of CITES, the aggregation of the expressions of willingness to pay generated there would result in the aggregation of preferences that are clearly in conflict when voting takes place on particular policies. The individuals who are most interested in the endangered species are interested in where the flows of value from the species will be channeled, not just that the stocks of the species will continue to exist.

This example is indicative of the general nature of the problem with which we are concerned. Just as the world's representatives to CITES are interested in the manner in which conservation policies channel the flows of values from the existence of an endangered species, we would suggest that the random individuals surveyed in a contingent valuation study would likewise assume that their expressions of willingness to pay would channel the flows of value in their intended directions. If they are to be able to aggregate these individual expressions into a meaningful expression of social preferences, the authors of such studies must investigate these underlying assumptions.

This discussion indicates that the economic model of the individual valuation of resource flows should be broadened in order to account for the values that individuals place on *flows* to others (McConnell, 1997). The literature on altruism indicates that many different forms of value may be present within an individual's utility function. The empirical work on existence values demonstrates to us that these more remote forms of value play important roles in individuals' own perceived welfare and policy choices. On the other hand, the results in this chapter equally indicate that the model that relies on the individual valuation of resource *stocks* should be reemphasized in order to avoid the conflation of very different and potentially conflicting flows of value. The focus on stock-related values confuses different flows and presents confusing models.

We will now explore these ideas by reference to a pair of studies undertaken by the authors. In the second section, we refer to the empirical literature on the valuation of stock-related values, and link this to dynamic forms of models on the valuation of stocks. In the third section, we illustrate the potential conflicts inherent in stated expressions of individual willingness to pay, by reference to a study on various management options that are potentially applicable to the black rhinoceros. In the fourth section, we illustrate the clear distinction between quantity-based and quality-based motivations in expressions of willingness to pay, by reference to a study on the provision of lands for use by the giant panda. In the fifth section, we discuss our results and in the final section we conclude.

Motives for Stock-Related Willingness to Pay

In this chapter, we will refer to a "motive" as being the source or underlying rationale for an expression of a willingness to pay for an enhanced stock of a natural resource. Since Krutilla's (1967) seminal work, the idea of an existence value has been interpreted as a form of stock (as opposed to flow) value. The concept of stock-related value has been analyzed in various parts of the economic literature. There is a range of possible reasons why an individual might register a willingness to pay for the maintenance of stocks of a given species. One part of the literature relates the concept to so-called *stock effects*, namely the impact of current stock levels on the costs of any future use of the species (see Neher, 1991). Another approach is to analyze the so-called *amenity value* of a given stock of a resource (Krautkraemer, 1989). This concept closely mirrors the nonconsumptive use value of a wildlife species – the flow of current value obtained without harvesting the resource. The *option values* of retaining stocks of a species have been analyzed as the choice concerning deferred harvesting or development (Arrow and Fisher, 1974). Another stock-related value that has been analyzed is the *bequest value* – the value of retaining some of the resource for passing on to the next decision-maker.

The important thing about all of these forms of stock value is that they are convertible to flows. The simplest example is that of a "stock effect" in a fishery. A fish in the sea has a stock value, but this value is simply representative of the present value of the future flow of reduced harvesting costs from leaving the fish unharvested now. So the current stock value for a fish merely represents the anticipated flow of values across time. Similarly, a person asked for his or her willingness to pay for a stock of (say) giant pandas or black rhinoceroses will be considering the flows of future value potentially generated from an enhanced stock.

In the case of the fishery, an individual's stock-related value will depend not only on how much value an enhanced stock will generate. It will also depend on the expectation concerning who will capture that value. If the individual fisherman expects to channel that flow of value to his own purposes (via private property rights in the fishery, for example), then the stock value will be the full present value of the future flow. If the individual fisherman expects that others will influence the future of that flow, then the stock value will be discounted for others' uses. Similarly, the person asked for his or her willingness to pay for an enhanced stock of an endangered species will be making assumptions about the channeling of those future flows of value.

In short, stock-related values for natural resources such as endangered species are often being considered within the framework of nonrival, nonexclusive public goods. Still, the flows from these stocks of natural resources often take the form of rival and exclusive goods, as in the case of the fishery. That is, individuals will be as interested in the manner in which enhanced flows will be channeled as they will be in the enhancement of the stocks themselves. If this is the case, then the manner in which flows are channeled from enhanced stocks is crucial to their correct valuation.

We have investigated these considerations in a pair of case studies concerning the giant panda and the black rhinoceros. We have investigated not just the willingness to pay for enhanced stocks of these endangered species, but also the motives for doing so. The motivations for the statement of a positive existence value for an enhanced stock of these species cover a wide and complex range; however, we believe that each may be converted into a statement about the channeling of a flow of value from that stock. So, for example, some of the various motivations underlying the statement of a positive stock-related value for a wildlife species might include the following:

Conservation motives

- *The option retention motive.* The importance of providing stocks of the species, in the belief that this will provide for the individual's own unplanned but possible future use, given that circumstances change to make this desirable. This is essentially channeling a flow of value from the species toward an unanticipated but possible "state of the world" in which the individual finds him- or herself in a situation to appreciate that flow.
- *The bequest motive.* The importance of providing stocks of the species, in the belief that these will provide for the betterment of future generations of human societies. This may be viewed as the channeling of flows of values toward future descendants of the respondent, and possibly even toward individuals about whom future descendants care.

Vicarious enjoyment (welfare) motives

- *Altruism.* The importance of providing stocks of the species, in the belief that other individuals than yourself are able to enjoy experiencing the species. This may be viewed as the channeling of flows of values toward others in the current generation about whom the respondent cares.
- *Animal welfare motive.* The importance of providing stocks of the species, in the belief that some will experience an enjoyable style of life or existence. This may be viewed as the channeling of resources toward the species and vicarious enjoyment of the animals' satisfaction.

Demonstration motives

- *The group identification motive.* The importance of providing stocks of the species in the course of identifying yourself as a member of a class or group of individuals who hold a particular set of beliefs. This may be viewed as channeling flows of value toward a group defined by a set of values or beliefs.

- *The belief specification motive.* The importance of providing stocks of the species in the course of specifying the beliefs that are derived from being a member of a particular group of individuals. This also may be viewed as channeling flows of value toward a group defined by a set of beliefs.

We have listed these motivations in this order to indicate those which we find to be most clearly based on the individual's own attempt to channel the flow of value from the species toward him- or herself. Conservation motives for stock values are those for which the individual sees no immediate or probable own use for the flows from the enhanced stock, but is willing to provide for the eventuality that one may exist. Welfare motives for stock values are those where the individual receives vicarious enjoyment from channeling flows toward clearly specified beneficiaries. Demonstration motives are those in which the individual is using the stock to assert a general principle regarding the channeling of values, preferring allocations toward certain groups, of which the individual perceives him- or herself to be a member. Thus each of the above rationales for stock-related values may also be viewed as an expression of interest in the channeling of the flows of values from those stocks, with the range representing increasing levels of abstraction in the designation of beneficiaries.

We will now turn to our case studies on the black rhinoceros and giant panda, in the third and fourth sections, respectively. These case studies were devised to assess the extent to which stock-related willingness to pay derive from interests in the channeling of future flows of values. We will describe the studies in the following two sections, while in the fifth section we will discuss our findings from these studies.

WTP for the Black Rhino Conservation Program – Conflicts in Conservation

The first study examined the WTP of the UK population for the conservation of the black rhinoceros in the country of Namibia (Swanson et al., 1998). In this case there was little likelihood that any of the respondents would ever actually experience the animals in question, but the respondents were still allowed to express their preferences for the continued existence of this highly endangered species via the vehicle of a fund established in the UK to pay for a conservation program in Namibia. The object of the exercise was to investigate the motivations for individuals' willingness to pay into such a fund, by observing the interaction between that WTP and the management programs to which the species was subjected. Would welfare-motivated individuals withdraw their support for management programs that enhanced conservation incentives by means that reduced the quality of life for the rhinoceros population?

The proposed conservation program in Namibia was entitled the Black Rhino Conservation Program (BRCP). Its aim was stated as follows: "to protect the existing Namibian black rhino population of 670 animals and to promote its increase to a minimum viable population of 2000, within the next 25 years." This would be achieved through the creation of heavily guarded rhino reserves within Namibia. Respondents were made aware of the total cost of the BRCP (estimated at 1 million US dollars per annum by Namibian officials) and of the fact that a current shortfall exists that would prevent the adoption of the proposed set of protective measures. Two possible ways of raising funds are: (i) the establishment of a UK Black Rhino Fund, that would be supported mainly by an environmental tax surcharge from UK taxpayers; and (ii) the establishment of a set of management programs that would develop various uses for the Namibian black rhinos, in order to generate amounts of money to sustain their conservation efforts in part.

There was then a presentation on the available black rhino *management options*: entry fees, live animal sales, sales of horns, de-horning, darting safaris, and trophy hunting (see figure 14.1). Attention was called to the fact that some of these

---

Option A – increase in entry fees

- Photographic safaris, viewing of animals in the wild.
- *Reduce international contribution from tax by 6 percent.*

Option B – sales of live rhinos

- A small number of animals (e.g., 6 of 670) can be sold each year on a long-term basis.
- *Reduce international contribution from tax by 10 percent.*

Option C – sales of stockpiled horns*

- Existing stockpiled horns may be marketed in a controlled trade setting.
- *Reduce international contribution from tax by 17 percent.*

Option D – de-horning operations*

- Safe procedure: shooting adult rhinos with tranquillizer guns and then sawing off their horns. Rhino horn re-grows: a horn is replaced in about 10 years.
- Harvested horns could be sold in a controlled trade setup (e.g., 83 of 670 rhinos).
- *Reduce international contribution from tax by 14 percent.*

Option E – darting safaris

- Tourist-hunters shoot rhinos with tranquillizer guns.
- Annual demand: around 10 hunts.
- *Reduce IDC by 4 percent.*

Option F – trophy hunting

- Tourist-hunters shoot and kill adult black rhinos.
- In small numbers (e.g., 3 of 670 rhinos) and in a controlled way. This would not endanger the survival of rhino populations.
- *Reduce international contribution from tax by 9 percent.*

Figure 14.1 Management options for black rhinos – a summary.

options (those marked with an asterisk) would only be available if legal trade of rhino products was to be allowed. It was explained that if all of the management options were adopted, 60 percent of the necessary funds for the BRCP would be collected. The remaining 40 percent would still have to come from international contributions to a government-organized fund (the Black Rhino Trust Fund).

Immediately after this, the group was presented with the valuation questions. The first WTP question asked for individual WTP for the full BRCP, when all the management options previously described were being used to help finance it. The format was open-ended and the payment vehicle was a one-time-only tax surcharge. In the second WTP question, hunting was deleted as an option to finance the BRCP. Respondents were asked for their new WTP to support the BRCP without the hunting of the rhinos. This question was designed to elicit the differential WTP for a preferred lifestyle for the species. The third elicitation question asked for WTP when all of the options that implied legal trade were deleted (sales of stockpiled horns, de-horning operations, darting safaris, *and* trophy hunting). This is basically the status quo – the only possible way to generate funds domestically from rhino conservation is through increased entry fees or live sales. The question was designed to assess the differential WTP for the removal of additional forms of intervention into the animal's lifestyle (de-horning and darting). Clearly, it also allowed the respondent to indicate any WTP for the continued ban on the rhino horn trade.

On average, respondents were willing to pay between £5 and £12.67 (depending on whether the median or the mean is used to summarize the data) for the full management Black Rhino Conservation Program, as a one-time-only contribution (see table 14.1).[2] As mentioned before, this program includes management options such as trophy hunting, de-horning operations, darting safaris, sales of stockpiled horns, sales of live rhinos, and increases in entry fees in wildlife parks.

The second issue that was investigated concerned the impact of varying management regimes on the values offered in support of the Namibian BRCP. As

Table 14.1 The value oⱼ ᴛhe BRCP, in pounds sterling

Total (N = 381)	WTP for the full BRCP	WTP for the BRCP with no hunting	WTP for the BRCP with no legal trade options
Mean	12.67	15.18	13.68
Median	5	10	5

[2] Given the presence of some outliers in the data – some very large bids – the median WTP is significantly lower than the mean. The median WTP provides a conservative estimate of the true value that could arguably be used for policy purposes.

mentioned above, respondents were offered the opportunity to pay for the set of management options that they preferred by means of registering different bids for different management packages. Specifically, they were given the opportunity to indicate their WTP for the full BRCP and also for the same package less trophy hunting. Then they were afforded the possibility of stating a WTP for the conservation of the rhino within a management program that disallowed almost all uses (trophy hunting, de-horning, darting safaris, and the trading of the horn), allowing only park fees and live sales.

Our first hypothesis concerned the potential conflict between welfare and conservation interests. These conflicts could be identified in various ways. If welfare concerns predominated over a general interest in conservation, the full BRCP would be the set of management options that would receive the lowest WTP, because it entailed the most intrusive set of management programs (all six) while generating the most conservation funding. Conversely, the status quo scenario, the less intrusive one that disallowed almost all commercial usage of rhino products, would yield the highest values. In addition, given the general public's dislike for sport hunting, it was anticipated that the elimination of rhino hunting would generate a significantly higher WTP than the full BRCP. Moreover, if welfare effects are strong, the elimination of further intrusive regimes (de-horning operations and darting safaris), and the denial of the commercial trade as well as sport hunting, might increase the WTP over that registered for the *full BRCP minus sport hunting*. Hence, it is interesting to investigate how the subtraction of further intrusive programs affects the nonuse value over the amount registered for the "nonhunting program" (that is, the full BRCP less trophy hunting). All of these comparisons are relevant to the determination of whether "welfare effects" or "conservation effects" predominate in the case of the commercial use of the rhino horn.

The first finding to report is that, as expected, there is a substantial WTP for a management regime that is devoid of all forms of sport hunting. If the BRCP does not include trophy hunting as a possible option to raise funds for rhino preservation, the mean WTP is £15.18 (table 14.1), which indicates that, on average, respondents are willing to pay an extra £2.51 to avoid trophy hunting of black rhinos (table 14.2). This difference is statistically significant both according to the Student's *t*-test of paired comparisons and the paired-rank Wilcoxon nonparametric test (table 14.3). The preferred measure of average WTP also indicates this difference in stark fashion: the median WTP doubles from £5 to £10 with the elimination of the use of the rhino for sport hunting.

Table 14.2 The values of several components of the BRCP, in pounds sterling

Total (N = 381)	Value of legal trade options minus hunting	Value of hunting	Value of all legal trade options
Mean	1.50	−2.51	−1.01
Median	0	0	0

Table 14.3 Results of hypothesis tests on the influence of several management options on the WTP for black rhino conservation

Estimated value	Null hypothesis	*t*-statistic decision	Wilcoxon test decision
Avoiding trophy hunting	$WTP_h - WTP_{fp} = 0$	Reject	Reject
Avoiding all legal trade options	$WTP_{lt} - WTP_{fp} = 0$	Cannot reject	Cannot reject
Legal trade options minus hunting	$WTP_h - WTP_{lt} = 0$	Reject	Reject

Note: all tests are two-sided and all decisions on H_0 are at the 95 percent level.

Next, the potential conflict between nonuse values and the use of the products that the black rhinoceros can generate was evaluated. Specifically, the survey groups were queried on the sensitivity of their WTP to the commercial usage of the horn of the black rhinoceros; that is, the regimes that implied the existence of a legal trade for rhino horn – sales of stockpiled horns, de-horning operations, darting safaris, and trophy hunting. Returning to table 14.1, the mean WTP for the BRCP without these options – the status quo scenario – is £13.68, an increase of about one pound over the full BRCP. This leads to the conclusion that respondents are not against having this set of options included in the program; that is, there is no perceived conflict between the nonuse value that the respondents are expressing and the use values derived from rhino horn trade. These two forms of value appear to be aggregative.

Further insights into the nature of respondents' preferences are possible from a closer look at the results. The status quo scenario, in which the BRCP excludes all regimes that imply the legal trade in rhino horn, rules out trophy hunting which, as was already seen, respondents dislike and has a negative value of £2.51 (respondents are willing to pay that sum to avoid it). Given that the value of the complete set of options that imply a commercial use of the horn is –£1.01 (respondents are willing to pay that sum to avoid it), it can be inferred that the value of the options that involve legal trade but not hunting is positive and equal to £1.50 (see table 14.2 and figure 14.2, "WTP distribution"), with this amount being statistically different from zero, as displayed in table 14.3. That is, respondents clearly are not giving a negative welfare-based valuation to some management options, such as de-horning and darting, while they are to others that are similar in intrusiveness, such as trophy hunting. Therefore, it may be concluded that there is a clear conflict between use and nonuse values in the case of trophy hunting but not in the case of the other uses (darting, de-horning, commercial uses, and live sales).

Figure 14.2 illustrates and summarizes the arguments presented in this section. The mean nonuse value for the existence of black rhinos lies somewhere within the range of £12.67 to £15.18 per UK household (or between £5 and £10 if the median is used), depending upon the lifestyle afforded to the animal in

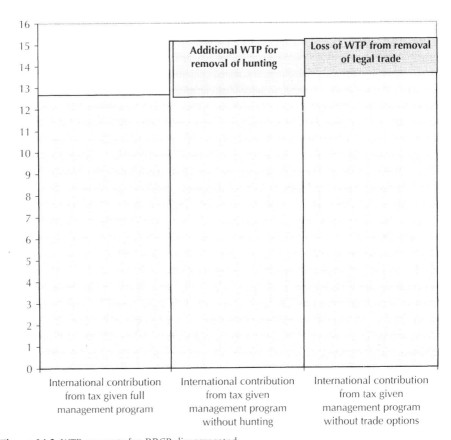

Figure 14.2 WTP amounts for BRCP disaggregated.

that jurisdiction. There is a mean positive WTP in support of both the removal of sport hunting from the BRCP (about £2.51) and of the inclusion of the rhino horn trade (about £1.50).[3]

How is it possible to explain these results? As discussed in the introduction, it was our hypothesis that the group would be driven by two different motivations when considering the policy options that they would like to support: an animal welfare-based motivation and a conservation-based motivation. Their welfare motivation would drive their WTP upward with the withdrawal of each additional intrusive option. However, a conservation-based motive would drive them to withdraw WTP when it was felt that the best set of options is not being afforded to the species. (This would be the case if the respondent believed that the absence of an optimal use-based conservation policy in regard to the rhino might reduce the prospects of receiving nonuse values deriving from motives

---

[3] It should be noted that this result was not unanimous. About 20 per cent of the respondents increased their WTP when the incremental commercial policies were excluded, and 42 percent of respondents decreased their WTP in the face of their elimination. On balance, the change in policies produced a significant downward shift in WTP.

other than animal welfare; for example, the person might feel that the possibility of making a bequest to future generations would be jeopardized by virtue of an imperfect management program.) These two different motives would be running in opposite directions in regards to the consideration of the various use values. We expected to find that there would be an additional (marginal) cost associated with the allowance of each additional intrusive option.

Instead, the survey results indicate that in order to maximize the nonuse values from rhinos, the most successful formula seems to be the banning of options that involve an element of enjoyment in the use of rhinos (hunting and darting) while allowing all other commercial uses of the animal, such as the sale of stockpiled horns and darting and de-horning operations. Interestingly, it does not appear that there is any additional withdrawal of support associated with intrusive management options other than those associated with sporting activity.

In short, it appears that our initial hypothesis, namely that the only apparent conflict between use and nonuse values concerned those constituencies with welfare motivations, was incorrect. This conflict exists to a very minor extent in the context of an endangered species such as the black rhinoceros, and it does not impact significantly upon the potential for all of the various use and nonuse values to be aggregated in the pursuit of conservation. However, there is an unanticipated source of conflict that has very significant implications for the aggregability of use and nonuse values. This is the conflict between those who enjoy specific forms of wildlife uses (namely sport hunters) and those who receive disutility from their enjoyment. This might be termed a sort of *vicarious disutility*; that is, one group values a loss of a flow of utility to another. This phenomenon was unanticipated, but appears to be the strongest source of conflict within the conservation community.

Paying for Panda Reserves – Motivating Existence Value

The black rhino study was followed by another pursuing some of the same issues in the context of the giant panda reserves in Sichuan, China. The giant panda is one of the most easily recognized and best-known species on Earth, and its ability to generate funds is well-documented, on both a use and a nonuse basis (Kontoleon and Swanson, 2000). In this study, we attempted to pursue the questions concerning conservation and welfare interests further: Why are individuals willing to pay for giant panda conservation? This was investigated in a survey of foreign tourists visiting the sights of China.

In this survey, a proposed conservation program for pandas was introduced: the Wolong Panda Conservation Program, which aimed to increase and maintain the panda population in the Wolong Panda Reserve to the minimum viable population of 500 animals. Respondents were informed that there were two possible scenarios for this program: the first involved removing pandas from the existing zoo cages in the breeding center of Wolong (the baseline situation) into pens which would provide 0.5 hectares per panda (a total of 250 hectares); while the second scenario would offer the 500 pandas the space of 400 hectares, in effect offering the pandas the entire reserve (200,000 hectares in total). It was made clear that the latter program would require more funds, since more land would have to be purchased for each panda. Thus the valuation questions were designed to assess the benefits from the enhancement of panda welfare as the results of the acquisition of more land.

The study attempted to separate between the respondents on the basis of already existing attitudes regarding conservation, animal welfare, and group membership. The attitudinal variables examined sought to reveal the existence of possible *animal welfare motives* (preferences for enhancing animal lifestyle) and demonstration motives (preferences for conformism to group norms). The variables that were used to proxy welfare motivations were indicators such as support for free-range products and antipathy to testing products on animals. The variables that were used to proxy demonstration motive were indicators such as the willingness to be seen wearing fur, or membership of a wildlife conservation organization.

Furthermore, respondents were asked directly about the consequences of possible panda extinction that concerned them the most. The question was designed to allow the respondent to declare expressly his or her motivations for wanting to conserve the panda – ranging from *pure conservation* motives (conservation of the gene pool for use that would maintain current and future flows to the respondents and to future generations) to *pure welfare* motives (conservation of species would enhance the flows to the species in the form of enhanced welfare). About 24 percent indicated that they were most concerned about the panda's own welfare and rights, while about 10 percent were most concerned only with impacts on the potential usefulness of the panda – others ranged between the two polar positions.

A final set of questions considered respondents' attitudes toward the proposed Wolong Panda Conservation Program. These were asked after information on pandas was given, and this exercise followed the valuation section. Nearly half of the respondents believed the program to be able to achieve the desired objectives, while the others claimed some level of distrust of government conservation programs.

As explained above, respondents were presented with a panda conservation program that would maintain the minimum viable population of pandas. They were then asked to state their WTP for the two different possible scenarios that this conservation program might offer: one involving the implementation of the program by placing giant pandas in pens, which would provide 0.5 hectares per panda; and the other by placing the pandas in the reserve, which would allow 400 hectares per panda. The assumption implicit in the valuation exercise is that additional land per panda increases its welfare or improves its lifestyle, but is likely to provide little additional panda conservation value. How would the WTP of the survey group vary between the two options?

Figures 14.3 and 14.4 show the distribution of additional WTP for each scenario. As expected, the WTP value varies inversely with the number of respondents reporting it. The results shows a large percentage of zero WTP for the pen option (27 percent), while only 8.43 percent of the sample were unwilling to

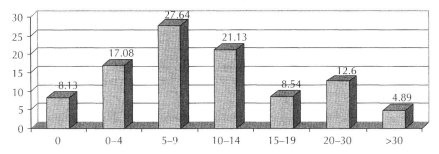

Figure 14.3 The distribution of WTP for the "reserve" scenario.

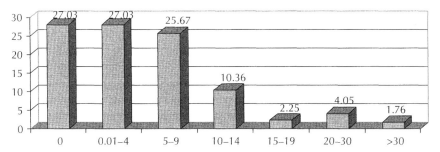

Figure 14.4 The distribution of WTP for the "pens" scenario.

Table 14.4 The mean and median of WTP for the scenarios, in US dollars

Total ($N = 303$)	WTP, pens	WTP, reserve
Mean	4.75	11.35
Median	4	10

state a positive additional value for "purchasing" the reserve for panda conservation. The survey revealed a mean positive WTP for the pen option of $4.75 per respondent, and a mean WTP for the reserve option of $11.35 (table 14.4).

What factors explain the amounts being bid for the pens and the reserves? Table 14.5 describes the explanatory variables used in the regression analysis. Table 14.6 presents the results from the estimated valuation functions. In both regressions, the coefficient for income is positive and significant, which is in accordance with economic theory. The expressed motive by the respondents (WCMOTIVE) is negative and significant in both equations, which indicates that the additional WTP for panda land is based more in welfare than conservation motives. That is, those individuals who are claiming welfare rather than conservation motives are willing to pay significantly more to provide more lands for pandas.

The other thing to note about the two different scenarios is the different set of significant variables. Both scenarios were significantly related to income levels, knowledge about pandas, and unwillingness to offend conservation/welfare group

Table 14.5 Variables used in regression analysis

Variable	Description
WTPpens	WTP amount, in US dollars, for scenario with pens
WTPreserve	WTP amount, in US dollars, for scenario with reserve
INCOME	Monthly disposable income, in US dollars
KNOWLEDGE	An index of the level of knowledge about pandas (1–3)
PROGRAM	An index of the level of trust, acceptability, and belief in the programs proposed (1–5)
WCMOTIVE	A weighted conservation motive for preserving pandas (0–3)
WELFARE	An index of general animal welfare/animal lifestyle values held by respondents (1–5)
DEMONSTRATION	Willingness to wear a fur coat if given one (1–5)
DPET	A dummy for owning a pet (1,0), as an index for empathy toward animals (1–5)

Table 14.6 Valuation functions for the "pens" and "reserve" scenarios

	Dependent variable			
	WTP/hectare under the "pens" scenario		WTP/hectare under the "reserve" scenario	
Variable	Coefficient	t-ratio	Coefficient	t-ratio
INCOME	2.82E–06***	3.256	4.67E–09***	3.235
KNOWLEDGE	0.009883**	2.188	1.32E–05*	1.713
PROGRAM	0.0017	0.639	9.55E–06**	2.118
WCMOTIVE	−0.01558***	−3.852	−4.43E–05***	−6.531
WELFARE	0.001558	0.407	1.01E–05*	1.557
DEMONSTRATION	0.006132**	2.362	1.04E–05**	2.349
DPET	0.001773	0.323	1.65E–05**	1.781
(Constant)	−0.02312	−1.171	−3.10E−05	−0.931
E(WTP/hectare)	0.021		0.000057	
R^2	0.17		0.31	
N	198		218	

Notes: *** significant at the 1 percent level; ** significant at the 5 percent level; * significant at the 10 percent level; corrected for heteroskedasticity.

interests. Only the reserve scenario bids were significantly related to: (i) belief in the conservation program; (ii) demonstrable welfare attitudes of the bidder; and (iii) demonstrated animal empathy (pet ownership).

The latter two are the important differences. Belief in the program merely separates between those who do and those who do not trust the payment vehicle – the panda conservation program. However, the significance of an individual's ability to empathize with animals and demonstrable welfare interests (purchases of free-range products and cruelty-free cosmetics, for example) indicates that the "existence value" in this case is more likely a flow of welfare to the species itself. Individuals are willing to pay more in order to influence the possibility that the individual animals will themselves acquire an increased flow of welfare-improving goods and services.

Conclusion – Explaining the Positive WTP for Stock-Related Values

The framing of the various motivations for conserving wildlife stocks as we have done above makes it clear that a positive statement of a "nonuse value" can mean many different things. In the studies conducted regarding the black rhino and the giant panda, we found evidence of many distinct and conflicting motives. For example, the respondents themselves provided an indicator of their *relative motivations* in the panda reserve study, and a relatively greater interest in animal welfare was significantly related to increased WTP. The conflicts between conservation and welfare motivations are even more apparent in the black rhino study, in which most respondents refused to provide more rhinos if they were to be used for the enjoyment of hunters. Increased stocks of the environmental asset are not valued in and of themselves – it depends on who will receive the benefits of these stocks.

We believe that individuals are willing to pay for enhanced stocks of a species because they believe that enhanced stocks correlate with an enhanced flow of goods and services to some other beneficiary (other individuals or groups, future generations, or the animals themselves). If the benefactor was able to provide for these flows to these groups directly, then the stock-related value would not be seen to exist. Positive statements of stock-related values, in this framework, act as surrogates for flows that are unable to be arranged otherwise. Payments for enhanced stocks then act as very crude instruments for the channeling of flows of goods and services in the desired direction.

This approach to existence value provides some explanation for the results acquired in the two studies discussed above. The respondents in the black rhino survey were providing an expression of their willingness to pay to channel flows of goods and services *away from hunters*. The respondents in the panda survey were providing an expression of their interest to channel flows of goods and services *toward pandas themselves*. Clearly, they are interested in much more than the simple provision of higher stock levels of the two species – they are crucially interested in the manner in which these stocks will benefit various groups.

What is the motivation underlying a positive WTP for increased stocks of an environmental asset? The answer to that question is as multifarious as the number of possible beneficiaries from the increased stocks. It is clear that individuals do value increased flows of welfare to those other than themselves, and it is not always apparent how they might attempt to channel these flows to others. So-called existence values, that are based on expressed WTP, are probably perceived by most respondents as very crude tools for influencing flows to others about whom they are concerned: future generations, other people, and even the animals themselves.

This does not denigrate the meaningfulness of an existence value, but it indicates the potential difficulties with rendering it objective and comparable. Depending on the species and its circumstances, the flows that its stocks might generate could be perceived to move in very different directions (hunters, tourists, and future users). Depending on the identity of the potential beneficiary, and the perception of the benefactor, the enhanced stock might generate a very different WTP (as in the case of the animal welfarist who is unwilling to provide rhinos for the hunter but willing to provide lands for the pandas themselves). The concept of any stock-related value should be tied closely to the perceived beneficiaries of those stocks.

More important, the concept of existence value and the now-ample evidence of its own existence should be taken as solid evidence of the substantial role that interconnectedness plays within an individual's utility function. Each of these bids is the best evidence of the impact that the welfare of others has on a given individual's own perceived welfare.

References

Arrow, K. J. and Fisher, A. C. 1974: Environmental preservation, uncertainty, and irreversibility. *Quarterly Journal of Economics*, 88(2), 312–19.

Barbier, E. B., Burgess, J. C., Swanson, T. M., and Pearce, D. W. 1990: *Elephants, Economics and Ivory*. London: Earthscan.

Kontoleon, A. and Swanson, T. 2000: Optimal panda ecotourism: a contingent valuation study of ecotourism development in the Wolong Panda Reserve in Sichuan, China. In D. W. Pearce (ed.), *Environmental Benefit Estimation: Developing World*. Cheltenham, UK: Edward Elgar.

Krautkraemer, J. A. 1989: Price expectations, ore quality selection, and the supply of a nonrenewable resource. *Journal of Environmental Economics and Management*, 16(3), 253–67.

Krutilla, J. V. 1967: Conservation reconsidered. *American Economic Review*, 57(4), 777–86.

Loomis, J. 1988: Broadening the concept and measurement of existence values. *Northeastern Journal of Agricultural Resource Economics*, 18, 20–9.

McConnell, K. E. 1997: Does altruism undermine existence value? *Journal of Environmental Economics and Management*, 32, 22–37.

Neher, P. 1991: *Natural Resource Economics: Conservation and Exploitation*. Cambridge: Cambridge University Press.

Swanson, T., Mourato, S., Swierzbinski, J., and Kontoleon, A. 1998: Conflicts in conservation: aggregating total economic values. Paper presented at the World Congress of Environmental and Resource Economists, 24–27 June 1998, Venice, Italy.

15

Environmental and Ethical Dimensions of the Provision of a Basic Need: Water and Sanitation Services in East Africa

Nick Johnstone, John Thompson, Munguti Katui-Katua, Mark Mujwajuzi, James Tumwine, Elizabeth Wood, and Ina Porras

In this chapter, we will explore issues related to the use of environmental resources by households in developing countries. The focus is on the use of water for drinking and cooking in East Africa. The use of sanitation facilities is also discussed, since it is closely related to use of water for other purposes. The chapter will explore the links between the environmental and ethical aspects of the provision of water and sanitation. The environmental aspects arise from the fact that water sources are often nonexcludable, and that water and sanitation provision generates, and are affected by, negative environmental and health externalities. The ethical aspects arise from the fact that society may attach value to an individual household's water consumption and access to sanitation facilities above and beyond the private household's (income-constrained) demand.[1]

The use of water resources by households for direct consumption presents us with a different problem than the amenity uses of environmental resources that form the usual subjects of discussion in the economics of environmental resources in OECD countries (see Dasgupta, 1997). The problems related to direct use of environmental resources in developing countries are distinct for two reasons. First, the links between the private consumptive uses of the environment and its associated public environmental benefits are often more complex than for amenity uses. The public environmental good is not something to be

---

[1] Water and sanitation services also have public attributes arising from the fact that the delivery of some kinds of services has natural monopoly characteristics, with economies of scale being very important. These issues will only be raised parenthetically.

cosseted and conserved, but is itself a private economic good (if not always a commodity) to be consumed. Second, unlike many other environmental resources, access to adequate water and sanitation provision is a public concern not only because of the more traditional concerns of nonexcludability and environmental externalities, but also because such access is a precondition for full participation in society, and even survival. It is a basic need and, as with all basic needs, society attaches a value to personal consumption patterns, even in the absence of negative environmental externalities and nonexcludability of resource use.

Thus, while negative environmental externalities and nonexcludability of resources are important in all societies, the real implications for households in different societies may differ greatly. In particular, it will be argued that since access to adequate and affordable water and sanitation facilities is fundamental to human health and the attainment of a reasonable standard of living in poorer countries, the existence of water-related negative externalities and nonexcludable water resources has very different implications than for some other environmental problems. Environmental market failures associated with such "basic needs" in poorer countries raise very different concerns than environmental market failures associated with amenity uses of the environment in richer countries.

In order to shed light on these links, this study reports on some of the results of an extensive survey of water consumption and environmental health in East Africa (Tanzania, Kenya, and Uganda).[2] Drawing on the survey, we will review the determinants of household access to "piped" water and sanitation services, since it is through piped services that some households are able to avoid some (but not all) of the problems associated with environmental externalities and nonexcludability. In addition, we will review the environmental and social implications of not having access to piped services. In particular, it will be argued that nonaccess to piped water services often results in reliance upon sources which are costly, far-removed, or polluted. And, finally, it will be argued that in order to ensure that households have access to the water and sanitation services necessary to meet their basic needs, greater use will have to be made of mechanisms that are emerging between the cracks of formal systems, many of which have failed because they were too ambitious. By overestimating demands that could be met, basic needs have been left unmet.

The second section will review the environmental and ethical aspects of the public nature of water and sanitation provision in more detail. The third section will report on the determinants of water consumption rates and access to piped

[2] The project was a collaborative research effort between researchers at the International Institute for Environment and Development (UK), the African Medical and Research Foundation (Kenya), Community Management and Training Services (Kenya), the Institute of Resource Assessment at the University of Dar es Salaam (Tanzania), the Uganda Community-Based Health Care Association (Uganda), and the Makerere University Medical School (Uganda). The study was a follow-up to the original "Drawers of Water" (DOW) study (see White, Bradley, and White, 1972); and financial support for "Drawers of Water II" was provided by the UK Department for International Development, The Dutch Ministry of Foreign Affairs, The Swedish International Development Cooperation Agency, and the Regional Office for East and Southern Africa of the Rockefeller Foundation.

water and sanitation facilities. The fourth section will review the implications of not having access to piped water services in terms of financial costs of water, inconvenience (distance and time), costs of collection, and health costs (diarrhea rates). The fifth section will provide a brief discussion of policy implications.

The "Public" Aspects of Water and Sanitation Service Provision

As noted above, environmental market failures are endemic to the provision of water and sanitation services. While it must be recognized that many sources of water are excludable, and have been for a considerable time, in other cases it may be prohibitively costly (or technically unfeasible) to restrict access in any way. In such cases, water use will be unregulated and the source is an open access resource. In other cases, water resources may have been held in common, with custom and tradition determining access effectively and efficiently. This is still true in many regions of developing countries. However, in some areas changes in social conditions or demographic change may have corroded this web of ties, resulting in conditions of nonexcludability (see Dasgupta, 1997).

Full or partial nonexcludability results in an excessive level of consumption. Households will have no incentive to internalize the additional costs of water consumption on other users as scarcity increases. This is particularly important for ground water. However, it also affects some surface waters, wells, and even public standpipes. If network water is not priced appropriately – as is the case in the majority of systems in developing countries – even piped supplies of water will be treated as an open access resource by users.

The difficulties involved in restricting access to water resources have contributed to decreasing availability of water resources for many households. At the global level, current trends indicate that the level of per capita available water resources is likely to continue to fall for the foreseeable future, with an estimated 250 million people living in areas under high water stress by 2020 (see OECD, 2001).[3] Some of the worst-affected areas are in sub-Saharan Africa. Of the three countries surveyed, Kenya faces the highest degree of water stress, but resources in some regions in all three countries are constrained.[4]

The low quality of water upon which households depend is often an equally pressing concern, with high incidences of a variety of waterborne and water-washed diseases. Negative environmental externalities associated with use of inadequate sanitation services are often very important contributors, with both surface and ground water affected. With an estimated 1.1 billion households in developing countries in 2000 not having access to an "improved" drinking water

---

[3] Water stress is considered high when the ratio of withdrawals (minus waste water returns) to renewable resources exceeds 0.4.

[4] According to the World Resources Institute (2000), annual withdrawals are 10 percent of water resources in Kenya, but only 1 percent for Uganda and Tanzania. However, national figures are of little practical relevance.

supply, and 2.4 billion households not having access to "improved" sanitation facilities, the problem is clearly pressing (WHO/UNICEF, 2000).[5]

The health consequences are considerable. According to the most recent Global Water Supply and Sanitation Assessment (WHO/UNICEF, 2000), there are 4 billion cases of diarrhea each year with 2.2 million deaths, most of which are children under the age of five. Intestinal worms such as roundworm and hookworm infect large proportions of the population of the developing world. Depending upon the severity of the infection, they can lead to malnutrition and retarded growth. A total of 6 million people are blind from trachoma. Other health concerns related to water and sanitation include schistosomiasis, cholera, and polio (see Hardoy, Mitlin, and Satterthwaite, 1992). In many cases, the adverse health effects of low water quality and inadequate water quantity are synergistic. The incidence of many of these diseases can be reduced through hygiene behavior, including the use of adequate amounts of water for washing (see Esrey, 1996).

The effects of many of those diseases listed above are borne by the wider community and not just by the household directly affected. Households may well recognize the adverse health effects of these diseases on their own family members and (if they can afford to do so) adjust their WSS provision patterns accordingly. However, they may not consider the external benefits of their own improved health to the health of the wider community. For instance, a household might choose to use a simple pit latrine, which is perfectly sanitary in terms of immediate environmental consequences. However, depending upon the soil conditions, it may result in externalities by contaminating the ground water supply of the community. Even if the household itself draws water from this supply, there will still tend to be excess contamination, since the household's cost of avoiding this contamination is likely to be greater than the household's expected benefit from better-quality ground water arising from their own efforts.

Thus, water and sanitation have strong "public" environmental attributes: water resources are often nonexcludable; use of inadequate sanitation facilities can result in negative environmental and health externalities; and consumption of water of poor quality (or in inadequate quantity) can generate negative health externalities. At the same time, water and sanitation are also necessities. In strict economic terms, this is reflected in the fact that estimated income elasticities for water demand are consistently less than one.[6] However, a much more fundamental case is also often made, with many arguing that access to adequate water supply and sanitation facilities is a "basic need." This is a controversial area, with the term itself being a subject of intense debate (see Sen, 1985).

At its core, the notion of a basic need draws upon Berlin's (1969) distinction between negative and positive freedom, with some goods being preconditions

---

[5] According to WHO/UNICEF criteria, an "improved" water supply includes household connections, public standpipes, boreholes, protected wells and springs, and rainwater. "Improved" sanitation includes public sewer connections, septic systems, pour-flush latrines, simple pit latrines, and ventilated improved latrines.

[6] Bahl and Linn (1992) review a number of country-level studies of water demand in developing countries and find estimated income elasticities ranging from 0.0 to 0.4. This is confirmed by cross-sectional evidence, indicating that the income-elasticity of water consumption is in the region of 0.3 (Anderson and Cavendish, 1993). Bhatia et al. (1995) review seven studies, reporting a range of estimated income elasticities from 0.15 to 0.78.

for "the ability of a person to function." The basket of goods and services that are considered to belong in this category will vary between societies and through time (see Helm, 1986). However, a strong case can be made for the inclusion of water and sanitation services in this category. Most fundamentally, a basic level of water consumption for drinking purposes is a precondition for survival itself (see Fass, 1993). Access to sanitation facilities, while less pressing in strict physiological terms, is nonetheless fundamental to meaningful participation in most societies. Thus, at one level, consumption is nondiscretionary. For instance, households do not "choose" to consume water for drinking and cooking purposes, but are physiologically required to do so.

Rawls (1971) labeled such goods as "primary goods," while Dasgupta (1986) uses the term "positive rights goods." The latter term underscores the fact that private consumption of water and private access to sanitation facilities both have public ethical dimensions. However, unlike the case of some other "goods" that can be classified as positive rights goods, consumers of water and sanitation services can affect each other's consumption possibilities and broader welfare directly. This is due to the nonexcludable nature of some water sources and the negative environmental and health externalities that exist. It is the joint existence of these two "public" elements (the environmental and the ethical) of water and sanitation services that has made public policy in the area such a fraught exercise.

The key point is that inadequate access to a basic need that is also potentially degradable and exhaustible can constrain the household's choices to such an extent that the choice itself can hardly be considered an exercise of freedom in any sense. In practice, household members are forced to choose between bearing costs in terms of potential ill-health, the use of extremely scarce financial resources (and thus other nondiscretionary consumption), or through large expenditures of time and effort. These issues will be explored in the fourth section. However, in order to provide the context for this discussion it is first necessary to review water consumption rates and levels of access to piped facilities.

Water Consumption Rates and Access to Network Water and Sanitation Facilities

The data and results presented in this and the following section are based upon a survey of over 1,015 households in East Africa. The survey was carried out in 1997 in the 33 East African sites studied in the original "Drawers of Water" (DOW I) study (White, Bradley, and White, 1972). Selection of these sites by the DOW I study team was "purposive," employing the available field assistants who returned to their home areas to carry out the study. In addition to returning to the original sites, the research methods that were adopted in 1997 were similar to those used for DOW I. The field assistants were university graduates who spoke the local languages, and were trained for two weeks. The training involved intensive workshops and fieldwork sessions, and provided an opportunity for the field assistants to familiarize themselves with the study's objectives and methodology.

Sample households in unpiped sites were selected using a grid of between 21 and 27 cells over an area of 8 square kilometers, using the same sampling method originally used by White, Bradley, and White (1972). A point within each cell was selected by using the coordinates of randomly selected numbers, and the household nearest to the point was chosen for interview. Piped sites were limited to the original urban areas studied in DOW I. Sampling in piped sited was quite different. Selected households in the piped sites were chosen by systematic random sampling, taking every tenth house and beginning at a number selected at random.

At each unpiped household, semistructured interviews were conducted and observations made. Data were collected on domestic water use, sociodemographic characteristics, the prevalence of diarrhea, the state and use of latrines, sources of water, and conditions of use. Wherever possible, reported water use was cross-checked by interviewing other respondents in the household and by observing the actual number of trips to the water source(s). Observations were carried out from 6 a.m. to 8 p.m. The actual amount of water used was measured by weighing it on a scale. Water used between 8 p.m. and 6 a.m. was estimated by interviewing household members. Information on environmental health, particularly on the prevalence of diarrhea, and the state and use of latrines, was obtained by interview and observation.

Water consumption rates

Not surprisingly, the survey revealed that water consumption rates differ markedly between piped and unpiped households. Mean water consumption for those with access is 46.62 liters per capita per day (lcd), but for unpiped households it is just 20.67 lcd. These figures are at the very low end of international consumption rates. For instance, a survey of urban and rural "recorded" domestic water consumption rates only recorded two countries (Bangladesh and Burma) with comparable rates (see Nickum and Easter, 1994). Moreover, the figure for unpiped households is only marginally higher than figures usually used as indicative of basic human requirements. For instance, the US Agency for International Development uses a guideline figure of 15–20 lcd for disaster relief projects involving "populations at risk" (USAID, 2000). A total of 230 households in the survey have average per capita consumption less than 15 lcd.

However, the figures for water consumption by use are more relevant for this discussion. Water consumption *per se* is not a basic need, but water consumption for some purposes is. For example, while the use of water for drinking and cooking may be considered as basic needs, it is clear that the use of water for other purposes – such as nonfood gardening, car washing, and swimming pools – may be a luxury good. This highlights the "instrumental" nature of water as a positive right good. In effect, it is really an input through which the positive right (a reasonable standard of health) can be realized.

Therefore, the survey data on water consumption was collected by use for drinking and cooking, personal hygiene, laundering and washing, toilet flushing, and gardening. Figure 15.1 provides data on water consumption by type of use for piped and unpiped households. The average consumption rate for drinking and cooking combined is approximately 4.5 lcd. However, 339 households had

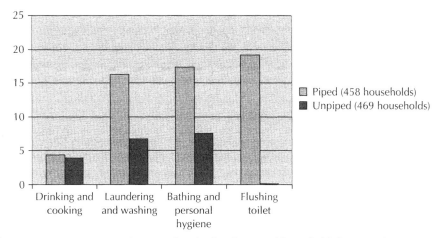

Figure 15.1 Water consumption rates for piped and unpiped households by type of use.

reported drinking water consumption rates less than the 3 lcd figure recommended in the aforementioned USAID guidelines.

In general, there was remarkably little variation in consumption rates across groups of households (see Thompson et al., 2000). The "nondiscretionary" nature of consumption for drinking and cooking is revealed by the similarity of the figures irrespective of whether or not the households have access to a piped connection. Indeed, much of the difference for the aggregate figures can be explained by flushing toilets, although the figure given (19 lcd) is based on a small subsample of only 104 households. Nonetheless, the differences in consumption rates for bathing and personal hygiene are large, and can influence the prevalence of negative health effects and externalities – for discussions of the relative importance of "water-washed" transmission routes, see Kolsky (1993) and Esrey (1996).

The nondiscretionary nature of water consumption for drinking and cooking is confirmed by econometric estimation, although the evidence provided is negative. While estimates of total (all uses) per capita water consumption consistently reveal significant coefficients of the expected sign,[7] per capita consumption of drinking and cooking water appears to be largely insensitive to economic and environmental conditions (for the full results of both estimations, see tables 15.2 and 15.3 of the appendix). The only factors that are statistically significant are those that are associated with household size and the proportion of children in the household. Presumably the former would reflect household economies of scale in the use of water for cooking and the latter would reflect the different "requirements" of children relative to adults. The economic variables (price and wealth) are not significant. More surprisingly, the variable that reflects whether or not a household has access to a piped connection is not statistically significant.

Not surprisingly, consumption of water for drinking and cooking appears to be nondiscretionary. Households consume approximately the same amount for these uses, irrespective of conditions. However, since the characteristics associated with

[7]   Although the price elasticity is very low.

alternative sources of water are very different, not having access to piped water can have significant financial, inconvenience, and health implications even if consumption levels are approximately the same. Thus, before reviewing some of these effects, the determinants of the likelihood of an individual household having access to a piped connection are first reviewed.

Access to piped facilities

Due to the discrete (qualitative) nature of the dependent variable, the determinants of a household having access to piped facilities were estimated using logit analysis. The dependent variable equals one if the household has a piped water connection, and zero if not. Explanatory variables included the household's country, the location (whether urban or rural, to reflect economies of density), the number of household members (to reflect household economies of scale in having a connection), a proxy for household wealth based upon the number of household members per room,[8] the number of years of education of the head of the household, and the number of years of residence of the household (to reflect the investment costs of obtaining a connection).

The model correctly predicted 82 percent of the cases. The estimated probabilities for each variable are presented in table 15.1 (the full results are presented in table 15.4 in the appendix). All of the coefficients except the dummy variable for Kenya and the estimated years of residency are of the expected sign and are statistically significant. The likelihood of a household having access to a piped connection increases by 5.1 percent for a 10 percent increase in the years of formal education of the head of the household. The dummy for location is also significant and large. Holding other factors constant, urban households are 53 percent more likely to have access to a piped water connection, presumably due to economies of density. The coefficient for household wealth is statistically significant, but not exceptionally large. A 10 percent increase in the wealth

Table 15.1 The likelihood of having access to "network" water

Variable	Weighted aggregate elasticity	
Kenya dummy	0.009	
Uganda dummy	−0.215	*
Urban dummy	0.532	*
Number in house	0.187	*
Years of education	0.509	*
Wealth proxy	0.336	*
Years in residence	−0.073	
Constant	−1.273	**

Note: *statistically significant at the 5 percent level; **statistically significant at the 10 percent level.

[8] Noncommensurable indices of relative wealth were used for piped and unpiped households, and so the proxy had to be used.

proxy (rooms per household member) results in a 3.4 percent increase in the probability of a given household having access to a piped connection.

A similar exercise was carried out for access to private network toilet facilities. In this case, the dependent variable was equal to one if the household had a flush toilet, as well as access to piped water facilities (inclusion of the latter serves as an additional check on the reliability of responses). The same explanatory variables as above are used, and the results are comparable, with considerable predictive power – there are just over 84 percent correct predictions. All but one of the variables (the dummy variable for Kenya) are statistically significant. However, the variable for years of residency is not of the expected sign (see table 15.4).

Thus, access to piped water and sanitation facilities is far from random. Wealthier, better-educated, urban, and large households are more likely to have piped connections. This is hardly surprising, and would be consistent with economic factors on both the demand and supply side. However, it does mean that it is often the poorer, less-educated, rural, and smaller households that are forced to make the most difficult choices about sources of service provision. The welfare implications of these choices are reviewed below.

It should be emphasized, however, that network water and sanitation facilities are by no means universally preferable (for an excellent discussion of the relative merits of different "on-site" sanitation facilities, see Mara, 1996). This is, of course, particularly true in rural locations where densities are lower, increasing the costs of network facilities and potentially reducing externalities from alternative systems (for a discussion, see Johnstone and Wood, 2000). Indeed, it will be argued in the final section that efforts to achieve universal access to network facilities may result in even lower levels of access to adequate facilities for poorer households.

The Welfare Implications of Not Having Access to Piped Facilities

The evidence presented in the third section indicated that households do not appear to bear the costs of nonaccess to piped facilities mainly in terms of reduced consumption of drinking and cooking water. This is not surprising, since – as   necessity – households are required to consume a minimum amount for survival. The choice that they face is not primarily about how much to consume but, rather, about their source of consumption. However, due to the very different implications of consumption from different sources, the costs of their choices manifest themselves in very different ways. In this section, we will review how households bear the costs of not having access to piped facilities through ill-health, financial costs, and/or inconvenience. All of these factors derive in large part from the public (in the environmental sense) nature of water and sanitation. Ill-health can be attributed in part to the existence of externalities and nonexcludability (increasing water scarcity). Inconvenience costs can also be partly attributed to scarcity. They can also be attributed to externalities that have affected more convenient sources. The same can be said of financial costs, although other factors are clearly also at play.

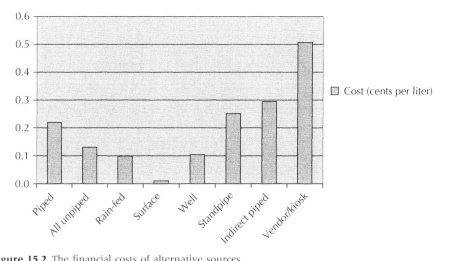

Figure 15.2 The financial costs of alternative sources.

Data on the incidence of diarrhea, the distance traveled to collect water, and the financial cost of water can be compared for households that opt (or are required) to use different classes of alternative source (rain-fed cisterns, surface waters, wells, indirect piped water from communal buildings or from neighbors, hydrants and standpipes, and vendors and kiosks). Most households without access to a piped connection will obtain their water from a number of these alternative sources. In some cases this will reflect the different uses to which the water is put. In other cases it may be a function of seasonal factors. Finally, in some cases it may reflect economic factors, as relative prices and other factors change. However, in the course of the survey, households were requested to designate a primary source. The largest group (219 households) used surface waters, followed by wells (113), vendors and kiosks (65), hydrants and stand-pipes (53), "other" sources (23), indirect access to piped through neighbors (20), and rain-fed water (11).

In figure 15.2, the financial effects of nonconnection are compared by type of "primary" source (a small number (12) of "outliers" had to be removed, since they were inconsistent with other figures reported for the same source from the same site). Not surprisingly, those who rely upon surface waters pay the least. Those who rely upon rainwater and wells are next, followed by standpipes and indirect piped (neighbors or communal building). Vendors and kiosks are by far the most expensive sources, with average costs that are more than double the price of more convenient direct "piped" water access. While the difference is significant, in a review of other studies Johnstone and Wood (2000) cite much higher ratios elsewhere.

In terms of "convenience" a rather different picture emerges, with vendors and kiosks being relatively close to the home (an average of just under 200 meters), while surface waters are further removed (over 400 meters, with 45 households at a distance of over 1 kilometer from their primary source; see

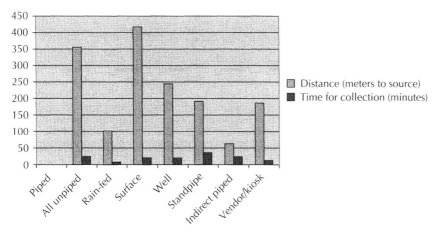

Figure 15.3 The "inconvenience" costs of alternative sources.

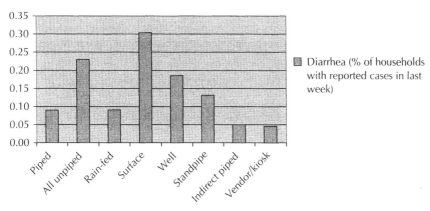

Figure 15.4 The health effects of alternative sources.

figure 15.3). Wells and standpipes are at an intermediate distance, while indirect piped access and rain-fed catchments are the closest of all. Not surprisingly, a similar picture emerges in terms of time for collection, although congestion at some types of sources (particularly standpipes) means that considerably more time is required than the distance would imply. Indeed, since these figures are equal to time required per individual trip, the 36 minutes required per trip on average for collection from a hydrant or standpipe means that a large proportion of the day can be spent collecting water. The "opportunity cost" of this time may dwarf any financial expenditures, and thus households clearly have incentives to trade off time against financial savings.

Perhaps more important are the health effects. Figure 15.4 compares source types with incidences of diarrhea per household. In this case, almost 30 percent of the households that relied upon surface water as their primary source reported at least one case of diarrhea in the previous week. Households that relied upon wells and standpipes were next, followed by those who relied upon rain-fed sources. Vendors and indirect piped access appeared to be the "safest" sources.

Indeed, they appeared to be somewhat safer than direct piped access. [Note that the characteristics of the sanitation facilities and hygiene behavior are also important determinants of diarrhea rates: see Kolsky (1993) and Esrey (1996).]

Two broad lessons can be drawn from this comparative discussion. First, unpiped households are generally worse off than piped households in terms of inconvenience and health effects. However, it is significant that the latter effect is not true for those who rely upon vendors. In terms of financial effects, there is some ambiguity, since many households have access to "free" sources (or are required to rely on "free" sources). Second, there is a tradeoff between alternative sources, with the less costly sources in financial terms having the highest inconvenience (standpipes and surface waters) and health (surface waters) costs. As noted above, all of these adverse welfare effects derive in part from the public environmental nature of the resource.

Thus, households appear to face a "tragic choice" between bearing the costs of nonaccess in financial terms (thus reducing already scarce disposable income) or bearing the costs in terms of inconvenience and health effects. Not surprisingly, it would appear that this choice is partly a function of relative wealth. Relatively poorer unconnected households tend to rely disproportionately upon "free sources," such as surface waters. Indeed, further (unreported) econometric evidence reveals that relative wealth is the most significant factor in determining the use of vendors or kiosks amongst unpiped households. A 10 percent increase in the ranking of relative wealth results in a 5.4 percent increase in the likelihood of using a vendor or kiosk rather than another source of water.

Conclusions

What are the implications of these results? As noted above, access to adequate water and sanitation is often defined as a basic need. While the precise basket of goods and services to be defined as basic needs is necessarily contextual, the case for the inclusion of water in this basket is very strong. This is particularly true in areas where the "public" environmental resource has been degraded. Households cannot rely upon surface and well water as an appropriate substitute. Indeed, many countries have codified the "right" to clean water. Thus a very different set of issues emerges relative to those associated with many other kinds of environmental resources, where issues of preference are to the fore. Some of these issues are ethical, insofar as a strong case can be made for the social value of personal consumption, even in the absence of externalities and nonexcludability.

This much is relatively uncontroversial. However, controversies do arise when this "right" is converted into practical policy priorities and measures. It is becoming increasingly clear that universal access to piped water and sanitation services is not a feasible policy objective in many countries, at least for the foreseeable future. Nor should universal access to standardized services be a policy objective, particularly in rural areas where costs can differ so much and where the implications of not having access can be so different. A focus on universal access to network facilities in many cases has retarded access to reasonable services for the majority of households (see Therkildsen, 1988). In many urban areas it has resulted in a dual system in which a small minority of households have access to high-quality services, and the large majority have to fend for themselves by whatever means possible (see Johnstone and Wood, 2000). In fact, in many sites in the survey households that previously had access to piped facilities no longer do so, due to widespread deterioration of infrastructure (for a comparison of the situation in 1967 and 1997, see Thompson et al., 2000).

The objective of universal access to standardized high-quality services has led to a situation in many countries in which a minority of households have access to (subsidized) water used in large part for nonessential purposes, while a majority of households are faced with a choice between a set of unsatisfactory alternative sources for water used to fulfill basic needs. As a consequence, the "basic needs" of many households are not met. Some households have access to low-cost, convenient, and safe water that is mainly used for "discretionary" uses, while other households are being forced to seek out more expensive, inconvenient, or unsafe alternative sources to satisfy their basic physiological and health requirements.

Clearly, classifying a good or a service as a "basic need" does not imply that there need be state provision of a homogeneous good to all households (see Dasgupta, 1993). Indeed, the good itself is merely an instrument through which the basic need is met. Rather than providing the good, the state can be a guarantor of its provision. In the context of water and sanitation, the public policy objective should be to ensure that households are not forced to make the "tragic" choices that they presently have to make in many regions. In the area of water and sanitation, this can mean a choice between using up scarce financial resources, expending vast amounts of time and effort, and risking their own health. Indeed, in participatory surveys undertaken in the study, a number of households emphasized that they did not see their decision about which alternative source to use as a choice at all (see Thompson et al., 2000).

From a public policy perspective, relaxing the constraints on this "choice" of water source means reducing the financial cost of vendor water and the inconvenience costs of public standpipes, or improving the quality of local water bodies. The latter is, of course, a desirable long-term objective, for both environmental and social reasons. Precarious environmental conditions (in terms of both scarcity and quality) are sharpening the ethical dilemmas associated with water provision. Indeed, in some areas improved sanitation facilities can be an effective means of increasing the availability of safe water sources. However, in order to ensure that households have access to affordable clean water in a reasonable time frame, the former two channels are of greater significance.

This is not the point at which to review the effectiveness of alternative policies and programs designed to provide poorer households with affordable access to adequate facilities. However, one area that is receiving increased attention is the increased use of small-scale private entrepreneurs and community-based organizations in the provision of both vended water and public standpipes. They are emerging between the cracks of failed delivery systems that involve much greater investment requirements. Indeed, in some cases they have even played a role in developing and managing small-scale infrastructure for service delivery (for broader discussions of these issues, see Silva et al., 1998; Solo, 1998).

Vended water has not usually been seen as part of the solution to bridging the deficit in access to affordable drinking water. On the basis of the financial costs cited above, this view may be warranted, since their cost would appear to indicate that they are an inefficient means of water delivery. However, in many cases the financial costs may be a reflection of rents arising from local monopolies. Alternatively, the high costs may reflect the risks associated with provision of a service that is not sanctioned. Where provision is competitive and legal, costs are often lower. Indeed, in recognition of their own capacity constraints, many public utilities have started to sell water to vendors for distribution in poorer neighborhoods (Solo, 1998). Given that this also allows for better control of water quality, formalization of the role of vendors may be an important step toward helping households to meet basic requirements for the foreseeable future.

Public standpipes are clearly also going to be important in helping households to meet such requirements. While initiatives pursued by development agencies, nongovernmental organizations, and community-based organizations have long focused on the provision of standpipes, there is room for institutional innovation in this area as well. In many cases, problems arise with maintenance, with many facilities falling into disrepair (see Thompson et al., 2000). This problem can be obviated by giving the managers of the standpipe a direct commercial interest in its upkeep. For instance, franchising of standpipes may be effective in some cases. It is also becoming more common to allow local community-based organizations to derive commercial benefits from upkeep (and even investment).

Finally, another more general conclusion emerges from the study, linking the environmental and ethical dimensions of water and sanitation service provision. As has been emphasized throughout, water is a basic need, but it is one that is drawn from a public environmental resource. The degree of excludability and the exposure to externalities is a function of the nature of service provision. It is highly likely that demand for the quality of "public" sources of water (surface waters and ground water) is dependent upon whether or not some households have access to "private" sources of water (piped networks). If households have access to treated water, they will not be particularly concerned about local environmental conditions. In effect, some households have managed (or feel that they have managed) to buy their way out of reliance upon scarce and degraded public water through private connections. A private good has been substituted for a public good.

Not surprisingly, the evidence indicates that it is the richer households that have been able to make this substitution. This means that demand for public water quality will be particularly low, due to the relatively low income "weight" of those households that continue to rely upon free (and often degraded and open access) water sources. For a period of time (until all local resources are completely exhausted and degraded), this may result in even greater divergence in the interests of the two groups of households. This will reduce political pressure to ensure that affordable, accessible, and clean water is available to all. The ethical and environmental dimensions of water and sanitation provision follow each other down in a vicious circle.

Appendix

Table 15.2 An OLS estimation of per capita total household water consumption (using the White correction procedure for heteroskedasticity)

Variable name	Estimated coefficient	Standard error	t-ratio, 875 d.f.	P-value	Partial correlation	Standard coefficient	Elasticity at means
PIPEDUM	12.590	1.920	6.557	0.000	0.216	0.2254	0.1904
KENYADUM	−3.735	1.830	−2.041	0.042	−0.069	−0.0630	−0.0401
UGANDUM	2.935	2.341	1.254	0.210	0.042	0.0519	0.0397
LOCATION	4.688	1.800	2.604	0.009	0.088	0.0826	0.0915
NUMHOUSE	−0.851	0.240	−3.546	0.000	−0.119	−0.1021	−0.1749
CHLDPROP	−16.533	4.108	−4.025	0.000	−0.135	−0.1388	−0.2001
FEMAPROP	9.248	3.734	2.477	0.013	0.083	0.0714	0.1642
WLTHPROX	4.796	2.164	2.217	0.027	0.075	0.1291	0.1472
EDUCATIO	0.919	0.219	4.198	0.000	0.141	0.1327	0.3163
USDLLITR	−1.107	3.066	−0.361	0.718	−0.012	−0.0130	−0.0062
GARDEN	18.703	4.977	3.758	0.000	0.126	0.2737	0.1063
CONSTANT	11.177	3.558	3.141	0.002	0.106	0.0000	0.3657

887 observations. Dependent variable = PCLITRES.
Using heteroskedasticity-consistent covariance matrix.
$R^2 = 0.3547$; R^2 adjusted = 0.3466.
Variance of the estimate − σ^2 = 507.18.
Standard error of the estimate − σ = 22.521.
Sum of squared errors − SSE = 0.44378E+06.
Mean of dependent variable = 30.560.
Log of the likelihood function = −4,015.06.

Table 15.3 An OLS estimation of per capita household drinking water consumption (using the White correction procedure for heteroskedasticity)

Variable name	Estimated coefficient	Standard error	t-ratio, 875 d.f.	P-value	Partial correlation	Standard coefficient	Elasticity at means
PIPEDUM	0.384	0.3021	1.271	0.204	0.051	0.0667	0.0308
KENYADUM	−0.749	0.3887	−1.927	0.054	−0.077	−0.1338	−0.0676
UGANDUM	−0.300	0.3591	−0.837	0.403	−0.034	−0.0556	−0.0407
OCATION	0.058	0.2456	0.240	0.810	0.010	0.0109	0.0081
NUMHOUSE	−0.155	0.0289	−5.389	0.000	−0.211	−0.1975	−0.2457
CHLDPROP	−2.345	0.6025	−3.890	0.000	−0.154	−0.2008	−0.2262
FEMAPROP	0.744	0.4651	1.601	0.110	0.064	0.0587	0.1013
EDUCATIO	−0.037	0.0306	−1.217	0.224	−0.049	−0.0547	−0.0958
WLTHPROX	−0.076	0.1903	−0.402	0.687	−0.016	−0.0219	−0.0173
USDLLITR	0.101	0.2880	0.352	0.725	0.014	0.0135	0.0048
CONSTANT	6.251	0.7306	8.556	0.000	0.324	0.0000	1.5484

634 observations. Dependent variable = PCDRINK.
Using heteroskedasticity-consistent covariance matrix.
$R^2 = 0.1077$; R^2 adjusted = 0.0934.
Variance of the estimate − σ^2 = 6.5868.
Standard error of the estimate − σ = 2.5665.
Sum of squared errors − SSE = 4,103.5.
Mean of dependent variable = 4.0374.
Log of the likelihood function = −1,491.62.

Table 15.4 A logit estimation of the likelihood of having access to "network" water services

Variable name	Estimated coefficient	Standard error	*t*-ratio	Elasticity at means	Weighted aggregate elasticity
KENYADUM	0.094	0.254	0.369	0.017	0.009
UGANDUM	−2.260	0.251	−8.984	−0.474	−0.215
LOCATION	3.129	0.268	11.644	1.031	0.532
NUMHOUSE	0.115	0.031	3.721	0.395	0.187
EDUCATIO	0.177	0.029	6.149	1.047	0.509
WLTHPROX	1.485	0.178	8.325	0.774	0.336
ESTIRESI	−0.063	0.043	−1.437	−0.158	−0.073
CONSTANT	−4.979	0.511	−9.742	−2.746	−1.273

Log-likelihood function = −388.91.
Log-likelihood (0) = −689.32.
Likelihood ratio test = 600.818 with 7 d.f.
Maddala *r*-square = 0.4533.
Cragg–Uhler *r*-square = 0.6045.
McFadden *r*-square = 0.4358:

- adjusted for degrees of freedom = 0.4318
- approximately *f*-distributed = 0.88280 with 7 and 8 d.f.

Chow *r*-square = 0.50083.

Prediction success table

		Actual	
		0	1
Predicted	0	420	88
	1	91	396

Number of right predictions = 816.
Percentage of right predictions = 0.820.
Sum of squared "residuals" = 124.08.
Weighted sum of squared "residuals" = 968.66.

Hensher–Johnson prediction success table

		Predicted choice		Observed count	Observed share
		0	1		
Actual choice	0	387.064	123.936	511	0.514
	1	123.936	360.064	484	0.486

Logit analysis, dependent variable = PIPEDUM.
995 total observations, 484 responded "yes."
Log of likelihood with constant term only = −689.32.
Binomial estimate = 0.4864.

Table 15.5 A logit estimation of the likelihood of having access to network sanitation facilities

Variable name	Estimated coefficient	Standard Error	*t*-ratio	Elasticity at means	Weighted aggregate elasticity
KENYADUM	−0.396	0.255	−1.556	−0.116	−0.041
UGANDUM	−2.006	0.256	−7.821	−0.614	−0.223
LOCATION	3.692	0.402	9.169	1.951	1.093
NUMHOUSE	0.129	0.034	3.793	0.683	0.262
EDUCATIO	0.273	0.036	7.545	2.534	1.113
WLTHPROX	1.576	0.192	8.181	1.289	0.492
ESTIRESI	−0.132	0.047	−2.784	−0.513	−0.173
CONSTANT	−7.868	0.733	−10.732	−6.650	−2.587

Log-likelihood function = −315.69.
Log-likelihood (0) = −590.26.
Likelihood ratio test = 549.131 with 7 d.f.
Maddala *r*-square = 0.4456.
Cragg–Uhler *r*-square = 0.6200.
McFadden *r*-square = 0.4651.
Adjusted for degrees of freedom = 0.4611.
Approximately *F*-distributed: 0.9939 with 7 and 8 d.f.
Chow *r*-square = 0.5109.

Prediction success table

		Actual	
		0	1
Predicted	0	560	84
	1	64	223

Number of right predictions = 783.
Percentage of right predictions = 0.8410.
Sum of squared "residuals" = 100.64.
Weighted sum of squared "residuals" = 853.29.

Ensher–Johnson prediction success table

		Predicted choice		Observed count	Share
		0	1		
Actual choice	0	532.198	100.802	624.000	0.670
	1	100.802	206.198	307.000	0.330

Logit analysis, dependent variable = FLUSH.
931 total observations, 307 responded "yes."
Log of likelihood with constant term only = −590.26.
Binomial estimate = 0.3298.

List of variables

PCLITRES	The per capita household consumption of water
PCDRINK	The per capita household consumption of water for drinking and cooking
PIPEDUM	A dummy equal to 1 if the household is piped
KENYADUM	A dummy equal to 1 if the household is located in Kenya
UGANDUM	A dummy equal to 1 if the household is located in Uganda
LOCATION	A dummy equal to 1 if the household is urban
EDUCATIO	The education level of the head of the household – to reflect health awareness
WLTHPROX	A proxy for wealth equal to number of rooms per household member
USDLLITR	The cost of water per liter
NUMHOUSE	The number of household members – to reflect economies of scale in water use
CHLDPROP	The proportion of children in household
FEMAPROP	The proportion of women in household
GARDEN	A dummy equal to 1 if the household used water for gardening
ESTIRESI	The estimated number of years in the residence – to reflect the investment horizon

References

Anderson, D. and Cavendish, W. 1993: Efficiency and substitution in pollution abatement. Discussion paper no. 186. Washington, DC: World Bank.

Bahl, R. W. and Linn, J. F. 1992: *Urban Public Finance in Developing Countries*. Oxford: Oxford University Press.

Berlin, I. 1969: *Four Essays on Liberty*. Oxford: Oxford University Press.

Bhatia, R., Cestti, R., and Winpenny, J. 1995: *Water Conservation and Reallocation: Best Practice in Improving Economic Efficiency and Environmental Quality*. New York: UNDP/World Bank Water and Sanitation Program.

Dasgupta, P. 1986: Positive freedom, markets and the welfare state. *Oxford Review of Economic Policy*, 2(2), 25–36.

Dasgupta, P. 1993: *An Inquiry into Well-Being and Destitution*. Oxford: Oxford University Press.

Dasgupta, P. 1997: Environmental and resource economics in the world of the poor. Discussion paper. Washington, DC: Resources for the Future.

Esrey, S. A. 1996: Water, waste and well-being: a multicountry study. *American Journal of Epidemiology*, 143, 608–23.

Fass, S. M. 1993: Water and poverty: implications for water planning. *Water Resources Research*, 29, 1975–81.

Hardoy, J., Mitlin, D., and Satterthwaite, D. 1992: *Environmental Problems in Third World Cities*. London: Earthscan.

Helm, D. 1986: The economic borders of the state. *Oxford Review of Economic Policy*, 2(2), i–xxiv.

Johnstone, N. and Wood, L. 2000: *Private Firms and Public Water: Realising Social and Environmental Objectives with Private Sector Participation in Water Supply and Sanitation*. Cheltenham, UK: Edward Elgar.

Kolsky, P. J. 1993: Water, sanitation and diarrhoea: the limits of understanding. *Transactions of the Royal Society of Tropical Medicine and Hygiene*, 87 (Suppl. 3), 43–6.

Mara, D. 1996: *Low-Cost Urban Sanitation*. New York: John Wiley.

Nickum, J. and Easter, W. 1994: *Metropolitan Water Use Conflicts in Asia and the Pacific*. Boulder, CO: Westview Press.

OECD 2001: *OECD Environmental Outlook*. Paris: OECD.

Rawls, J. A. 1971: *A Theory of Justice*. Cambridge, MA: The Belknap Press of Harvard University Press.

Sen, A. 1985: *Commodities and Capabilities*. Amsterdam: North-Holland.

Silva, G., Tynan, N., and Yesim, Y. 1998: Private participation in the water and sewerage sector – recent trends. *Public Policy for the Private Sector note no. 147*. Washington, DC: World Bank.

Solo, M. T. 1998: *Small Scale Entrepreneurs in the Urban Water and Sanitation Market*. Washington: UNDP/World Bank Water and Sanitation Programme, mimeo.

Therkildsen, O. 1988: *Watering White Elephants? Lessons from Donor-Funded Planning and Implementation of Rural Water Supplies in Tanzania*. Uppsala, Sweden: Centre for Development Research Publication 7, Scandinavian Institute of African Studies.

Thompson, J., Porras, I. T., Wood, E., Tumwine, J. K., Mujwahuzi, M. R., Katui-Katua, M., and Johnstone, N. 2000: Waiting at the tap: changes in urban water use in East Africa over three decades. *Environment and Urbanisation*, 12(2), 37–52.

USAID 2000: *Field Operations Guide: Version 3.0*. Washington, DC: USAID.

White, G., Bradley, D., and White, A. 1972: *Drawers of Water*. Chicago: University of Chicago Press.

WHO/UNICEF 2000: *Global Water Supply and Sanitation Assessment: 2000 Report*. Geneva: WHO/UNICEF Joint Monitoring Programme.

World Resources Institute 2000: *World Resources 2000–2001*. Washington, DC: WRI.

PART VI

Conclusions

CHAPTER 16

Economics, Ethics, and Environmental Policy

Daniel W. Bromley and Jouni Paavola

The foregoing chapters have demonstrated that research at the intersection of environmental economics and ethics focuses attention on how choices related to the environment are, should, or should not be made. It should be clear, as well, that the approaches of economics and ethics to environmental problems are not easily reconciled in the realms of either individual or collective choice. In this chapter, we present one synthesis of how collective choices about environmental policy can be understood in research at the intersection of economics and philosophy.[1] We hope that this discussion contributes to the debate in economics and philosophy about the nature of environmental policy.

Philosophers ask how people frame the environmental choices they face. Is one more item in their consumption bundle actually *traded off at the margin* for the protection of one more hectare of wilderness? Are pizza and ptarmigans just two *interchangeable objects* giving of "utility?" Economists conjure an indifference curve relating the two objects and find normative significance in the implied prices that reduce this choice to a monetary equivalent. Most philosophers would reject this exercise as conflating two distinct human endeavors. Philosophers might well agree that one *purchases a pizza* for the satisfaction it will bring. On the other hand, they often insist that one may make a *commitment to ptarmigans* for their own sake – despite the satisfaction that one might gain by making that commitment.

Philosophers have no difficulty in distinguishing between the two behaviors, yet the distinction escapes many economists. The usual response from an economist might well be that an individual cannot both eat her pizza and have the ptarmigans too, so choices must be made. Notice that this approach is concerned with choice more than it is with value. Economists avoid discussions of the *value*

---

[1] We shall use the term "collective choice" to denote policy action taken by parliaments or administrative agencies, or legal decrees issued by courts. The word "collective" therefore departs somewhat from conventional economic usage, where it tends to connote group behavior (more than one individual). We regard public policy as *collective action in liberation and restraint of individual action.*

of pizza and ptarmigans by casting the matter as a choice between more pizza and fewer ptarmigans, or between more ptarmigans and less pizza, insisting that their relative value is *revealed by the choices that people make*.

Moreover, many economists advocate the choice-theoretic models of the utility-maximizing individual as the correct way to formulate collective choice and public policy. This implies that collective choice is nothing more than ascertaining the rate at which the *aggregate* of individuals wishes to trade pizza for ptarmigans. On this tack, all human action – from shopping for pizza, to making collective choices in the legislature, to joining a community group concerned with neighborhood attributes – is seen as simply the pursuit of individual self-interest, animated by gains and losses, pleasure and pain, or benefits and costs. Small wonder that philosophers have a problem with economists' treatment of choice.

Economics has not always been thus. When hedonism was gaining ground in economics, Thorstein Veblen observed, in a famous and sarcastic quote, that:

> The hedonistic conception of man is that of a lighting calculator of pleasures and pains, who oscillates like a homogeneous globule of desire of happiness under the impulse of stimuli that shift him about . . . When the force of the impact is spent, he comes to rest, a self-contained globule of desire as before. . . . He is not the seat of a process of living, except in the sense that he is subject to a series of permutations enforced upon him by circumstances external and alien to him. (Veblen, 1990: 73–4)

Contemporary environmental economics largely incorporates the view of human behavior criticized by Veblen. This approach has a limited ability to accommodate the choice between pizza and ptarmigans in terms of *other-regarding behavior* – or commitment, as Sen (1977) calls it. Philosophers often reject hedonism as an explanation of, and a guide to, choices between disparate alternatives. This position partly explains their aversion to monetary valuation of the environment, and the use of thusly derived "values" in a benefit–cost analysis. Philosophers are not alone – many citizens react similarly. Choices about life, quality of life, and personal integrity are often characterized by moral judgments and commitment. Shared social norms demarcate the applicability of different *modes* of choice in different *domains* of choice by allowing, restricting, or rejecting the *commodity fiction* (Radin, 1996).[2]

The moral dimensions of environmental choice have been explored by Edwards (1986), Gregory and McDaniels (1987), Harris et al. (1989), Kahneman and Knetsch (1992), Opaluch and Segerson (1989), and Stevens et al. (1991). Stevens et al. (1991) studied the importance of the survival of different species in New England with the contingent valuation method. According to their results, the most important reason to support habitat restoration was the existence value attached to the species – even for species such as salmon that also have use value. A majority (79 percent) of their respondents agreed with the proposition

[2]   The idea of the commodity fiction originated with Karl Polanyi (1965).

that "all species of wildlife have a right to live independent of any benefit or harm to people" (Stevens et al., 1991: 396). Yet, when the survey offered an opportunity to pay to secure the preservation of the species, the majority of respondents refused.[3] The authors suggest that the respondents were ". . . either uncertain about their valuation, believed that wildlife should not be valued in dollar terms, or protested the donation payment vehicle. Moreover, most of those who would pay exhibited behavior that appears inconsistent with the neoclassical theory underlying the CVM" (1991: 399). Referring to Harper (1989) and Opaluch and Segerson (1989), the authors argue that their survey may have asked people to choose between ordinary goods (income) and a moral principle, which is likely to result in behavior that is inconsistent with the usual assumptions concerning preferences.

The approach followed by Stevens et al. in their valuation survey did not seem to fit the perceptions of their respondents. Since individuals actually form their views about environmental goods in the process of choosing (responding to the survey), there is little common understanding regarding how such issues *ought to* be framed and evaluated. Certainly the information offered to respondents influences their bids.[4] The respondents' moral commitment to respect the "right" of species to exist could explain their reluctance to assure that survival by exhibiting a willingness to pay for it. Economists have difficulty explaining this kind of behavior, because economics does not have a coherent and sound view of choice informed by moral commitment.

The widespread commitment of environmental economists to uncover what they regard as the "value" of environmental resources with contingent (or other) valuation methods is a logical consequence of the standard economic approach which regards prior knowledge of prices to be necessary for rational choice. The same view explains the commitment to benefit–cost analysis when deciding about environmental (and other) policies (Palmer, Oates, and Portney, 1995; Arrow et al., 1996). This form of *a priorism* is challenged by those who approach environmental policy from a philosophical viewpoint (Sagoff, 1998; Holland, this volume; Norton, this volume), as well as by those who take an empiricist approach (Porter and van der Linde, 1995). There are also compelling theoretical reasons to challenge the coherence of predicating collective choices over environmental policy on hypothetical prices for sundry parts of nature (Diamond and Hausman, 1994; Vatn and Bromley, 1994).

There is yet another reason to be skeptical of the idea that environmental policy is but individual hedonism writ large. The reason is that economic analysis treats individual choice as a *mechanical process*. The tastes and preferences of

---

[3] The survey portrayed a hypothetical reduction in public spending on wildlife restoration in New England. A private trust fund was presented as a way to compensate for the diminished public funding. The respondents were asked to offer hypothetical payments to this fund. The authors argue that some responses reveal a view according to which habitat restoration is a public matter and should not be handled by a private trust.

[4] Sagoff (1988) asks what is the right amount of information to be given in the survey experiment. That is, how much discussion, deliberation, and "learning" are acceptable?

individuals are alleged to exist and be known *a priori*. Hence, individual choice is modeled as nothing but a mechanical process of acting on those priors to respond to external stimuli in an arena – the market – whose *raison d'être* is to give effect to preexisting preferences. The standard view is that individuals come to a market not to discover what they want, but to get what they already *know* they want. Knowing their desires, individuals assess prices and quantities and choose a combination that will make them better off than any other combination. This view of choice is *mechanical* because there is nothing about choice that requires *explanation*.[5] The individual automatically chooses so as to maximize utility. The arguments deployed to "explain" choice are identical to those an engineer might us to "explain" indoor temperature. Economic agents produce utility-maximizing outcomes for the same reason that a thermostat produces an optimal room temperature: the "settings" of both define what is optimal and there is no scope for deviation from that optimum. Choice here is not explained, it is "justified" as being consistent with preferences. A justification is not an explanation.

To reiterate, economists regard choice as a process in which individuals take their tastes and preferences as given – and prices are the parameters against which tradeoffs occur at the margin. The variable in this formulation is utility: it is maximized when no reallocation of one's budget can improve it. Environmental economists apply this stylized story of individual choice to collective choice, and then seek to present the decision-maker with clear evidence of the benefits and costs of various alternatives so that the "optimal" choice can be taken. This approach is followed – and strongly advocated – on the ground that in the absence of adherence to the rigors of economic logic, it would be much too easy for politicians to do great harm.[6]

We now turn to a broader challenge to such *a priorism*. We shall argue that monetized benefits and costs are rarely useful – and certainly not decisive – when making collective choices over environmental policy. On the contrary, we will suggest that environmental policy is best understood as an exercise of *pure practical reason*.

Prices, Values, and Choice

The classical economics of Smith, Ricardo, and Marx focused on the problem of *value* in economic affairs. When neoclassical economics replaced the older school of thought, it turned its back on the problem of value and concentrated instead on the much less controversial problem of choice. This shift in emphasis was thought to enhance the status of economics as a science (Cooter and Rappaport, 1984). The interesting question is: How have economists managed to address choice while remaining silent on value? Is this not a logical contradiction?

---

[5]   See chapter 2 of this volume, by Alan Holland.
[6]   For an example of this view, see Palmer, Oates, and Portney (1995).

Economics has finessed the potential contradiction with the *deus ex machina* of *indifference*. Marginal utilitarians argued that we do not need to worry about the problem of value as long as we place all goods on the same plane and observe how consumers weigh incremental units of one good *vis-à-vis* another good (Lewin, 1996). Or, if we wish to be more general, we can observe how individuals weigh a unit of one good against all other goods by considering the expenditure on the good as against the remaining income that could be spent on any combination of all other goods.

In this model, prices do not reflect *value* but, rather, *indifference*. At the margin, where all choice is made, a consummated transaction indicates that the buyer and the seller have arrived at a point of perfect indifference. The buyer is indifferent between giving up money for the good in question or keeping the money – and the seller is in a comparable situation. This must, by definition, be the case – value is always relative to all other goods that might be bought with one's income. This reasoning also suggests that choice *cannot* be driven by *a priori* notions of value. Rather, choice is a process from which "value" emerges as a point of indifference between two or more possible choices. The individual chooses and "value" is the *resultant* of that choice, not its *cause* (antecedent). That is, when a transaction is finished, we may say tautologically that the price reflects value to the buyer and the seller under the relevant circumstances of the choice. But price is a reflection of value in only the most backhanded of ways: contrary to what we teach, *a price results from a particular choice rather than being the cause of that choice*.

This logic suggests that individual choice is rather different from how it is usually portrayed and modeled. Price must indeed be seen as the result of choice rather than as its cause. Moreover, the concept of preferences, the relationship of preferences to choice, and the explanation of choice must be reconsidered. The reason is simple. The conventional understanding of preferences does not allow us to explain choice. To say that choices reveal preferences does not *explain* but, rather, *justifies* action. Choices do not reveal preferences in the sense of indicating to us why an individual preferred α rather than β (Paavola, ch. 6, this volume). To say that choosing α instead of β offered more utility to the individual does not provide an *explanation* for why α was chosen over β. Utility is not a reason for choice when utility merely indicates the satisfaction level of individuals' preferences.

Explaining choices involves asking questions about the *reasons for* those choices. The first question to be asked is: "What counts as reasons for actions?" As above, preferences or utility do not count as reasons. Reasons relate to the consequences of choice or to what is *accomplished by a particular choice*. To reiterate, choice must be regarded as an individual's contemplation of plausible *reasons for action*, and then taking that action (choice) for which the individual can muster the best reasons. Reasons for action can be many. It may indeed be true that some choices entail nothing more than considering one more unit of something versus all other things that might otherwise be chosen. Other choices, however, may be predicated on – motivated by – other reasons, such as securing the preservation of an endangered species. Here it is not obvious that doing so involves tradeoffs.

Consider an individual who is contemplating the choice between apple pie and cherry pie. After some thought, she chooses the cherry pie. An economist could announce: (1) that she preferred the cherry pie to the apple pie; (2) that she obtained greater utility from the cherry pie than from the apple pie; or (3) that choosing the cherry pie was consistent with her preferences. However, these propositions do not *explain* her choice but merely provide *justifications* for it. We would not be helped either by declaring that she behaved *as if* she preferred the cherry pie. If a physician were to tell us that we were sweating *as if* we had a minor cold, when in fact we had contracted malaria, and were then to proceed to treat the wrong reason for the fever, that physician would be liable for malpractice.

Coherent explanation of choice in our pie example would be made more difficult if the agent were to announce that she preferred and wanted apple pie, but still chose the cherry pie because she feared that apples might contain pesticide residues. How should we describe and explain her choice? Did she choose cherry pie even though she did not prefer it? Or did she indicate that she ordinarily prefers apple pie but that today she preferred cherry pie? This reply is not reassuring if one believes that preferences are stable: the agent could be reminded that her preferences could not possibly change so quickly. But that would be unlikely to persuade her.

We can complicate the example further by indicating that the known odds of our subject suffering any ill effects from the pesticide residue on apples are less than one-half the known odds that the cherry pie might trigger a serious allergic reaction. How would we explain her refusal to change her choice in the light of this new information? What would be the reasons for her action? We know the *reasons for her action* when we can advance a true statement that contains a proposition as to *why* she chose the cherry pie rather than the apple pie. To say that one augmented her utility more than the other does not explain her reasons for action (Cooter and Rappoport, 1984).

We have offered two challenges to conventional choice models. First, we insisted that choices between disparate alternatives, such as pizza and ptarmigans, occupy two distinct mental realms. Agents usually choose pizza by considering only themselves, while choosing ptarmigans quite likely involves the consideration of others – either the ptarmigans or those agents who also care for them. Second, we insisted that choices must be seen as searching for the best reasons to undertake action α rather than action β. The explanation for choice is most certainly not the price, because price results from choice (through the intermediation of indifference with respect to all other possible choices). The *explanation* for a choice is found in its future consequences – having and/or enjoying α rather than β.[7]

We now turn to a detailed discussion of the process of searching for plausible reasons for action.

---

[7] We must allow choices based on rule-following. In this limiting case, undertaking an act or making a choice is itself the sought-after consequence.

Calculation and Sentiment

The rational decision-maker of contemporary economics is said to act only after engaging in careful calculation to determine which choice will maximize utility. When these calculations are not carried out, the action of the individual – whether habitual, or action predicated on emotion – is regarded as reactive or ill considered. This view privileges economics in that it celebrates the notion that only economics offers precise algorithms to "explain" all choices – which are properly characterized by the calculative form of rationality. We may notice the circularity in this idea. This circularity constitutes justificationism when economists define a particular method of choice as *rational* and therefore the proper way in which choices *ought* to be made. Moreover, there is no empirical evidence to support the universal applicability of the conventional view of choice. As Herbert Simon has said, "'Reasonable men' reach 'reasonable' conclusions in circumstances where they have no prospect of applying classical models of substantive rationality" (1978: 14).

We suggest that it is useful to understand rational action as predicated on *contemplation*. Contemplation consists in the deployment of alternative mental models to pending action. In the first instance, contemplation involves asking "Should I do what I have always done in this situation, or should I behave differently?" The first alternative entails another question: "Should I act without thinking about it?" It is common to believe that action predicated on habit is not rational action. However, to decide to do what one has always done can be the essence of rationality when information is costly to obtain and process.

If one decides to break with habit, then the problem becomes "What should I do?" There are two alternatives. The first is to choose to search for and process data to obtain the *decisive information* upon which action can be predicated. The second alternative is to choose a different choice algorithm: "I shall weigh a variety of considerations and I shall act when I have marshaled enough good reasons to discriminate between doing *X* and doing *Y*. Then I shall choose *X* or *Y*." We shall refer to the first mode of choice as *calculation*, while the second one shall be referred to as *sentiment*.

Let us first consider calculation. By calculation we mean a process of constituting the decisive predicates of action by computing "probabilities and payoffs." Standard economic explanations are based on such calculation: a rational agent considers alternative actions, calculates probabilities associated with them, identifies which action will most enhance her utility, and then chooses. Of course, conventional economics also allows for short-cuts if, for example, limited cognitive capacity forces the agent to stop short of a "full" computational exercise before choice. Agents also "learn" from their earlier choices, develop rules of thumb, and then apply these short-cuts in subsequent choice situations to reach *boundedly rational* decisions.

However, while ideas about limited cognitive capacity and bounded rationality are heuristically valuable and useful, they fail to rescue the idea that all action is predicated on calculation. That is, actions not predicated upon full calculation are not exhausted by those that are predicated upon *optimally incomplete* calculation. We insist that efforts to rescue the conventional understanding of rationality as being characterized exclusively by calculation are not successful. Many of the earlier chapters have given abundant evidence in this regard. Therefore we insist that economics must recognize a realm of reason that transcends mere calculation.

For example, the person considering which pie to eat had her *reasons for action*, yet those reasons seem at odds with what we usually understand of rational choice. "Ordinary" rationality would have her choose the (possibly) pesticide-laden apple pie rather than the allergy-inducing cherry pie, yet she did just the opposite. To invoke bounded rationality in the example suggests a strategy of redefining and justifying inexplicable choice as boundedly rational. Actually, her choice cannot be the product of – is not predicated upon – calculation. Nor is her choice of cherry pie the product of habit. We are left to conclude that her choice must belong to the realm of *sentiment*. What do we mean by sentiment?

Sentiment does not involve calculation, nor is it identical with reactive or habitual action. Sentiment is a ground for choice when agents consciously reject calculation and rely, instead, on reasoning that can be consequentialist (either utilitarian or nonutilitarian in nature), or deontological (emphasizing moral "right" regardless of implications). However, it is still action informed by contemplation – albeit over a range of reasons that cannot be calibrated and calculated so as to admit careful tradeoffs and thus calculation.[8] We suggest that sentiment is the application of reason to choice (action), where the reasons for choice are the result not of computation but of *due consideration*. Sentiment provides sufficient grounds for rational action if we mean by it action that is consistent with an agent's purposes.

We insist that all human action is a blend of calculation and sentiment. This can be seen in the current debate over genetically modified organisms (GMOs). The scientific community developing and advocating GMOs is convinced of their benefits and insists that they pose *no known risks*. Their critics dismiss these assurances as yet another illustration of technological optimism, and assert that there are no reliable means whereby the long-run risks can be assessed. The critics also reject *a priori* risk assessments of such technologies. For this, they often are referred to as Luddites or romantics, who refuse to be rational and to listen to the evidence of the scientific community. The proponents claim to be dealing with the facts, while suggesting that their opponents appeal to emotion. The advocates insist that public policy about GMOs should be guided by (their) "science" and not by political posturing (Feldman, Morris, and Hoisington, 2000). The realm of calculation is regarded as a rational basis for action, while the realm of sentiment is portrayed as the antithesis of rationality. We are reminded of Martin O'Connor's (ch. 12, this volume) statement that *the absence of proof of danger is not proof of the absence of danger*. Technological optimists are confused on the point.

Recognition of the complementary realms of calculation and of sentiment offers an opportunity to consider the economic and ethical dimensions of environmental policy in an integrated framework. Economics does not – and need

---

[8] Here our approach parallels that of Holland (ch. 2, this volume).

not – fight with ethics, or vice versa. Ethical considerations need not be viewed – or modeled – as orthogonal to economic calculations. Nor does one make a "tradeoff" between economic logic and ethical reasoning. Rather, each of us draws from two different realms of reason when contemplating particular actions. We deliberate in all choice situations (by the very necessity of making a choice) and in doing so we may look for reasons for action in the realm of calculation, or we may look for reasons in the realm of sentiment. Very often, our action – our choice – will be predicated upon a combination of the two.

This recognition of the two realms of reason allows us to understand policy initiatives for environmental protection that seem to bypass the normal insistence that all environmental legislation must pass a benefit–cost test before it can be regarded as "rational." One often sees economists insisting that a particular environmental action – new legislation – was a case of "politics trumping economic rationality." But public policy is simply collective action in liberation and restraint of individual action. When nations collectively decide – and that is precisely the *raison d'être* of parliaments in democratic states – that it is time to eliminate a particular environmental insult, say the use of chlorofluorocarbons, those engaged in collective action (parliamentarians and other interested parties) will draw from both realms to provide reasons for their actions.[9] In these settings we regard it as somewhat odd to denounce politicians because they have refused to accept the finding of some benefit–cost study and decided, instead, to base their action on some other grounds. It could be argued that their behavior is consistent with the "job description" of politicians. If we wished for them to be – and to act like – economists, then we should change their job description, and increase their salary.

We now turn to an alternative characterization of the policy process that admits of the relevance of realms of calculation and realms of sentiment.

On Practical Inference

The standard economic approach to environmental policy is predicated on consequentialist welfarism that places "... exclusive reliance on individual utilities to judge *social* goodness and right actions" (Sen, 1993: 521). We suggest, however, that collective choices about the environment – undertaken in parliaments, courts and in administrative agencies – are fundamentally different from the choices that individuals make on their own account. This suggests that welfarist approaches – given all of their problems at the level of individual choice – are even less pertinent for understanding and explaining collective action. Moreover, while individual choice is undertaken within an institutional framework that defines permitted fields of action, collective action is concerned precisely to modify that institutional framework.

Parliaments, courts, and administrative agencies are venues of collective action in which the institutional framework is modified to alter future social states.

---

[9]   This has been described as "making choices without prices without apologies" (Vatn and Bromley, 1994).

Despite arguments for the use of potential compensation tests, institutional change is fundamentally Pareto-noncomparable. As Coate recently put the matter, ". . . why should the fact that the gainers *could* compensate the losers make socially desirable the infliction of those losses?" (Coate, 2000: 438).[10] More important, arguments for compensation tests are profoundly circular. The potential compensation test cannot act as a welfarist guide for collective choices over public policy, because new policies are simply new institutional arrangements (new working rules) that alter endowments and future income streams across the population into the future. These institutional arrangements and the distributions of income and wealth they engender, generate Pareto-optimal outcomes that are and remain incomparable in Pareto terms (Bromley, 1990).

All public policies – and most assuredly environmental policies – simply ratify particular interests that seem, at the time, to merit protection by the adoption of particular new institutional arrangements. The collective ratification of certain interests, requisite institutional arrangements, and the outcomes they will engender remind us that truth is that which it is better, at the moment, to believe (Rorty, 1999). Is it now better to believe that polychlorinated biphenyls (PCBs) threaten public health? Is it now better to believe that chlorofluorocarbons (CFCs) pose a serious threat to atmospheric ozone? Is it now better to believe that northern spotted owls are more important than the harvesting of trees in which those birds nest? Is it now better to believe that breaching the dams on the Snake River is necessary to assure the survival of Columbia River salmon runs? Is it now better to believe that nutrient enrichment of the Baltic Sea must be stopped at all cost? Is it now better to believe that Finland and Sweden should join the European Union, while it is better to believe that Norway should not?

Each policy decision has been and will be informed by contemplation in the realms of both calculation *and* sentiment. The realm of calculation cannot be – and clearly is not – alone decisive in policy choices.[11] For some of the above policy decisions, a great effort was undertaken to generate data and calculations on the belief that this would lend a gloss of "rationality" to the choice. While some of these exercises may have been useful, others were of such dubious merit that they were discredited or ignored. At the end of the day, decisions have been (and will be) made on the basis of what, quite simply, seemed better to believe. Truth is that which seems prudent, at the moment, to believe.[12]

In democratic states, public policy can be regarded as an application of the syllogism of *practical inference* (von Wright, 1983).[13] A syllogism of practical inference

[10] The Scitovsky Paradox reminds us that ". . . compensation criteria can . . . simultaneously recommend introducing and removing the same policy change" (Coate, 2000: 438).
[11] Notice that we do not say that the realm of calculation *should* (or should not) be decisive. Our task here is to be descriptive of how policy is considered and implement, not how it *ought to be* implemented on economistic grounds.
[12] In the UK the long debate on – and the many benefit–cost studies of – the third London airport illustrate the futility of seeking to arrive at complex decisions by reductionist methods.
[13] The syllogism of practical inference applies to individual action as well, but we will limit our discussion to the realm of collective action.

brings together two kinds of premises. The first is the *volitional premise*, a proposition concerning an *end of action*. The volitional premise expresses a desired future outcome for the sake of which a particular action *must* be undertaken now. If the desire is to protect the Baltic Sea from further eutrophication, then particular actions are required now. If the desire is to preserve the salmon in the Columbia – Snake River system, then particular actions are required now. If the desire is to protect atmospheric ozone, then particular actions are required now. We see here the application of *prospective volition* – the human will in action, looking to the future, contemplating ways in which the future could and *should* unfold. If nothing is done, a particular future will unfold as an outcome of existing institutional arrangements. If this future is regarded as undesirable, then a change of existing institutional arrangements is instrumental in altering the future. This brings us to the *epistemic premise*.

The epistemic premise draws on scientific and traditional knowledge to offer a plausible guide for necessary action to realize the volitional premise. If atmosheric ozone is to be protected, then the epistemic premise is that chlorofluorocarbons must be eliminated from everyday use. If it is intended that the Baltic must be protected from further eutrophication, then the epistemic premise suggests that nitrogen and phosphorous loads must be reduced. If it is intended that Columbia River salmon must survive, then the epistemic premise reveals the plausible plan of action to realize that outcome.

To reiterate, the very essence of public policy starts with a consideration of the problematic nature of the status quo ante, and a consideration of desired future outcomes (the *volitional premise*). The *epistemic premise* – of the form "if X then Y" – connects the desired outcome (X) with the necessary action (Y) to achieve that outcome. The epistemic premise is both a prediction and a prescription. The epistemic premise *prescribes* what (Y) must be done in order to achieve the desired outcome (X), and it *predicts* that the desired outcome (X) will be realized if the action (Y) is undertaken. Policy-making as an exercise of practical inference does not require estimates of monetary benefits of these future states in order to contemplate the wisdom of the requisite institutional change that will generate those future states.

The conclusion of a syllogism of practical inference is referred to as a *practical necessity*. The conclusion points to the practical necessity of deploying the means (Y) implicated in the epistemic premise in order to attain the end (X) implicated in the volitional premise. The necessity of the conclusion of practical inference follows from the nature of the syllogism. The volitional premise is *not* of the form "X is desired if the benefits of X exceed the benefits of $\sim X$" (where "$\sim X$" means "not X"). Nor is the volitional premise of the form "X is desired if the benefits of X exceed the costs of Y." Rather, the volitional premise states what *must* be done. The discourse of parliaments, the deliberation of the courts, and the consideration of administrative agencies establish the relative merits of X and $\sim X$. If collective action were simply an instrumental exercise of calculating the benefits and costs of policy alternatives, then parliaments, courts, and administrative agencies could be replaced by councils of economic advisors or by central planners. Many environmental economists resist the prospect of making choices without prices, but this is a misplaced concern.[14] Democratic structures and processes exist for precisely this purpose.

There is another important difference between traditional policy analysis and practical inference. Benefit–cost analysis seeks to justify *future economic circumstances in terms of the present.* Consider a policy that entails costs for those of us now living and that yields benefits for unborn, such as a policy to reduce the emissions of greenhouse gases. The Earth's climate is getting warmer, and there is now plausible evidence that a good share of this warming is caused by human activity. The conventional approach would be to calculate the present value of future benefits and costs of reducing the emissions of greenhouse gases. If the net present value of the policy were found to be negative, then most economists would announce that it a mistake to control greenhouse gas emissions. Global warming would be regarded as a Pareto-irrelevant externality (Buchanan and Stubblebine, 1962).

This approach is claimed to assure us of policies that will maximize social welfare over time. But this cannot possibly be true. The approach discounts the utility of future persons, which cannot be defended on either efficiency or equity grounds (Ramsey, 1928; Broome, 1983; Chichilnisky, 1997). As Frank Ramsey once observed, the affinity for discounting results from a lack of imagination and is unethical (Bazelon and Smetters, 1999). Discounting monetary benefits and costs obscures the fact that we are discounting the life prospects of *future* persons. Namely, benefit–cost analysis considers the future in terms of the *internal* rate of time preference, which is used by living agents to contemplate deferring their consumption into the future. Yet we are not deciding about deferring our own consumption but, rather, are deciding about the nature of the environment to be inherited by yet unborn persons. The *external* rate of time preference would indicate the rate at which those of us now living choose to discount the utility of future persons. Intertemporal choices, in which we compare our present consumption or investment decisions against similar decisions of future persons, consist in interpersonal comparisons of utility. Intertemporal choice is *interpersonal* choice. The idea that we (the gainers) could compensate the future – the losers if we fail to take corrective action – is incoherent. To suppose that we can "compensate" them by not correcting environmental problems and by passing more wealth on to them – in their increasingly degraded environment – simply compounds the fallacy of exclusive reliance on the realm of calculation.

Contrary to conventional economic thinking, the political process – the means whereby societies make collective choices – brings the future into view. Individuals may contemplate their future in terms of the institutional framework within which they are embedded, and which defines for them possible domains of choice and the future their choices are likely to engender. However, individuals cannot change the institutional framework alone – only collective action in the parliaments and the courts can do that. When contemplating that institutional framework and its possible alteration, the future is all that matters. What good would it do to alter institutional arrangements looking to the past?

---

[14] See Vatn and Bromley (1994).

Thus a benefit–cost analysis of environmental policies that create costs today – higher carbon abatement costs, for example – while the benefits accrue to unborn persons, considers the future in *terms of the present*. It grants us a dictatorship over the environmental assets to be inherited by future persons in that it allows us to act in *our* own interest, not in the interest of future persons. It remains to be explained how this approach can be said to maximize social welfare over time. Benefit–cost analysis assures that the time stream of future net benefits is as large as possible to those of us now living and choosing. This means acting so that the future serves the present. Discounting serves those of us living today very well indeed, since it discourages environmental policies that impose upon us the inevitable costs of adjustment, while yielding benefits to future persons.

Recall that the intertemporal problem has been primarily debated in terms of appropriate discount rate – once again locating the problem exclusively in the realm of calculation. However, intertemporal choice is essentially one in which the realm of sentiment is pertinent. The fundamental question here is "What is right with respect to the future?"[15] Ordinary welfarism cannot answer that question. Welfarist answers might well be possible behind the Rawlsian Veil of Ignorance: What decision would be taken about a policy problem (such as global climate change) by risk-neutral agents who were ignorant of whether they would live today or 100 years from now? In the language of a super-fair game, intertemporal environmental policy would be predicated upon the idea of *no envy*. The decision should be such that no agent, upon learning *when* he or she would live (now or 100 years from now), would wish to trade places with any other agent living in any other time. Notice that the pertinent question is not what climate endowment those of us living now *prefer* to leave for future persons. Rather, the problem is how can the tyranny of time's arrow be solved in the interest of all present and future persons? The constitutional rule of "no envy" would address that problem.

The other alternative is to rely on the realm of sentiment in which there is a commitment based on other-regarding (or nonwelfarist) ethical premises to act so as to respect the interests of the unborn. This brings us back to the notion of prospective volition – the human will in action, considering the present in terms of the future. That is, what actions must be taken now in order that the future shall be better than the past and the present? This vision sees reasons for action running *from the future back to the present*. This vision of the policy problem requires the concept of *final cause* (which is also the volitional premise of practical inference):

> the "final cause" of an occurrence is an event in the future for the sake of which the occurrence takes place . . . things are explained by the purposes they serve. When we ask "why?" . . . we may mean either of two things. We may mean: "What purpose did this event serve?" or we may mean: "What earlier circumstances caused this event?" The answer to the former question is a teleological explanation or an explanation by final causes; the answer to the latter question is a mechanistic explanation. (Russell, 1945: 67)

---

[15]   See chapter 3 of this volume, by Bryan Norton.

The idea of final cause reveals that pollution is reduced not because it is economically efficient to do so, but because of a collective moral commitment. Chlorofluorocarbons were not prohibited by the Montreal Protocol because the net present value of doing so was found to be positive. Rather, they were banned because of a collective commitment to arrest the further destruction of atmospheric ozone (and to restore it). Finland and Sweden did not join the EU because the net present value of doing so was found to be positive. These nations joined because their citizenry contemplated a future in and out of the Union and concluded that being in was *probably* better than being out. The citizens of Norway made the opposite choice. These choices are precisely the result of contemplation in the realms of both calculation and sentiment.

Implications

The idea of environmental policy is just that – an idea. It is a mental model of how environmental policy *is* and *ought* to be made. Yet, when acted upon, ideas about environmental policy seriously matter because they hold very real consequences for those of us now living – and for the generations that will come after us. This is why ideas about economics and ethics applied to environmental policy merit serious analysis – a view that gave rise to this volume.

Philosophers and economists most assuredly approach environmental policy from different premises and angles. Most economists approach environmental policy with the efficiency conditions of Pareto optimality in mind. Their approach to environmental policy is *instrumental*: the calculations and comparisons – the weighing of benefits and costs – are a means (an instrument) to find the "optimal" policy. While the optimality judgments of economics are entirely self-referential, many economists still imagine that their policy prescriptions are value neutral, and that the solutions *predicated on their calculations* represent unimpeachable truths to which reasonable people *ought to* subscribe. What "rational" person could possibly favor a policy that is pronounced by economists as failing an efficiency test?

Economists consider their approach to environmental policy to be legitimate because it is *instrumental to the Pareto condition*. The normative appeal of the Pareto test rests on two central value judgments: (1) only individuals count; and (2) a policy that makes one person better off, without at the same time making another worse off is, by definition, an improvement in social welfare. If one accepts these value judgments then one must also accept the Pareto test with its efficiency entailments as the rational guide to environmental policy. The irony, of course, is that while economists insist that they address means (instruments) rather than ends (outcomes), the Pareto test constitutes nothing if not policy outcomes. If people clamor for cleaner air, economic analysis will tell us if it is really "rational" to do so and what the "optimal" level of air quality would be. If the economist finds that the aggregate willingness to pay for cleaner air is less than the costs of making the air cleaner, then the economist will declare that it is "irrational" to wish for cleaner air. The air is optimally foul – or foul air is simply a Pareto-irrelevant externality.

In contrast, philosophers usually see environmental policy as constitutive rather than instrumental: environmental policy it is about who we are and who we seek to become. For a philosopher, the problem is *how* do we arrive at judgments concerning what is thought to be good? Quine quotes Neurath to say that the philosopher's task is akin to that of "a mariner who must rebuild his ship in the open sea" (Quine, 1980: 70). The philosopher takes up those very things that economists seek to avoid through the latter's embrace of instrumental reason. Neurath's mariner cannot afford the luxury of instrumentalism.

Can philosophers and economists narrow their profound methodological and epistemological differences? We do not believe so, but this is not necessarily reason for despair. The convergence of ideas and approaches is neither a necessary nor sufficient sign of intellectual progress, but the capability to reflect upon and to justify the chosen metaphysical positions arguably is. We believe that the discussion on the metaphysical differences between economic and philosophical inquiries into environmental policy can and must continue. Doing so will foster, we believe, reflection and learning both among economists and philosophers. The need for this enduring conversation between economists and philosophers is underlined by the fact that shared ideas about environmental policy have very concrete consequences. The conversations would need to focus on first principles of the two disciplines, as well as on the truth claims that each can offer. Only in this way will two disciplines so central to human progress make relevant contributions to the public discourse about who we are and who we are to become.

References

Arrow, K. J., Cropper, M. L., Eads, G. C., Hahn, R. W., Lave, L. B., Noll, R. G., Portney, P. R., Russell, M., Schmalensee, R., Smith, V. K., and Stavins R. N. 1996: Is there a role for benefit–cost analysis in environmental, health, and safety regulation? *Science*, 272 (12 April), 221–2.

Bazelon, C. and Smetters, K. 1999: Discounting inside the Washington, DC beltway. *Journal of Economic Perspectives*, 13(4), 213–28.

Bromley, D. W. 1990: The ideology of efficiency: searching for a theory of policy analysis. *Journal of Environmental Economics and Management*, 19(1), 86–107.

Broome, J. 1983: Discounting the future. *Philosophy and Public Affairs*, 23(2), 128–56.

Buchanan, J. and Stubblebine, W. C. 1962: Externality. *Economica*, 29, 371–84.

Chichilnisky, G. 1997: The costs and benefits of benefit–cost analysis. *Environment and Development Economics*, 2, 202–6.

Coate, S. 2000: An efficiency approach to the evaluation of policy changes. *The Economic Journal*, 110 (April), 437–55.

Cooter, R. and Rappoport, P. 1984: Were the ordinalists wrong about welfare economics? *Journal of Economic Literature*, 22 (June), 507–30.

Diamond, P. A. and Hausman, J. A. 1994: Contingent valuation: is some number better than no number? *Journal of Economic Perspectives*, 8(4), 45–64.

Edwards, S. F. 1986: Ethical preferences and the assessment of existence values: does the neoclassical model fit? *Northeastern Journal of Agricultural and Resource Economics*, 15, 145–50.

Feldman, M. P., Morris, M. L., and Hoisington, D. 2000: Why all the controversy? *Choices* (first quarter), 8–12.

Gregory, R. and McDaniels, T. 1987: Valuing environmental losses: what promise does the right measure hold? *Policy Science*, 20, 11–26.

Harper, C. R. 1989: Rational roots of irrational behavior. *Northeastern Journal of Agricultural and Resource Economics*, 18, 96–7.

Harris, C., Driver, B. L., and McLaughlin W. J. 1989: Improving the contingent valuation method: a psychological perspective. *Journal of Environmental Economics and Management*, 17, 213–29.

Kahneman, D. and Knetsch, J. L. 1992: Valuing public goods: the purchase of moral satisfaction. *Journal of Environmental Economics and Management*, 22, 57–70.

Lewin, S. B. 1996: Economics and psychology: lessons for our own day from the early twentieth century. *Journal of Economic Perspectives*, 34, 1293–323.

Opaluch, J. J. and Segerson, K. 1989: Rational roots of "irrational" behavior: new theories of economic decision-making. *Northeastern Journal of Agricultural and Resource Economics*, 18, 81–95.

Palmer, K., Oates, W., and Portney, P. R. 1995: Tightening environmental standards: the benefit–cost or the no-cost paradigm? *Journal of Economic Perspectives*, 9(4), 119–32.

Polanyi, K. 1965: *The Great Transformation*. Boston: Beacon Press.

Porter, M. E. and van der Linde, C. 1995: Toward a new conception of the environment–competitiveness relationship. *Journal of Economic Perspectives*, 9(4), 97–118.

Quine, W. van Orman 1980: *From a Logical Point of View*. Cambridge, MA: Harvard University Press.

Radin, M. J. 1996: *Contested Commodities*. Cambridge, MA: Harvard University Press.

Ramsey, F. 1928: A mathematical theory of saving. *Economic Journal*, 38, 543–9.

Rorty, R. 1999: *Philosophy and Social Hope*. London: Penguin.

Russell, B. 1945: *A History of Western Philosophy*. New York: Simon and Schuster.

Sagoff, M. 1988: *The Economy of the Earth: Philosophy, Law and Environment*. Cambridge: Cambridge University Press.

Sagoff, M. 1998: Aggregation and deliberation in valuing environmental goods: a look beyond contingent pricing. *Ecological Economics*, 24, 193–213.

Sen, A. 1977: Rational fools: a critique of the behavioral foundations of economic theory. *Philosophy and Public Affairs*, 6, 317–44.

Sen, A. 1993: Markets and freedoms: achievements and limitations of the market mechanism in promoting individual freedoms. *Oxford Economic Papers*, 45, 519–41.

Simon, H. A. 1978: Rationality as process and as product of thought. *American Economic Review*, 68(2), 1–16.

Stevens, T. H., Echevenia, J., Glass, R. J., Hager, T., and More, T. A. 1991: Measuring the existence value of wildlife: what do CVM estimates really show? *Land Economics*, 67, 390–400.

Vatn, A. and Bromley, D. W. 1994: Choices without prices without apologies. *Journal of Environmental Economics and Management*, 26, 129–48.

Veblen, T. 1990: Why is economics not an evolutionary science? In *The Place of Science in Modern Civilization*. New Brunswick, NJ: Transaction Publishers (originally published by Viking Press, New York, 1919).

von Wright, G. H. 1983: *Practical Reason*. Ithaca, NY: Cornell University Press.

Index